Incredible Nikola Tesla

Mysteries Solved

By Steve Preston

1

1st Edition

Table of Contents

Table of Contents

4

Who Was Nikola?

Of anyone in the world, the most mysterious of men would have to be Nikola Tesla. He counted his peas, couldn't stand to look at pearls, talked to invisible people, developed an earthquake machine, had an unimaginable photographic memory, developed the first radio and wireless remote-controlled vehicle, fought a deadly game with Thomas Edison, and while all this was going on, he solved many of the mysteries of our universe. Most know him for inventing AC electric power, and from his attempt at bringing free electricity to the entire world, but there is so much more to this icon of invention, insight, and wonder and we are go to find it.

This book will open your eyes to some of the miracles of Tesla and fill in some blanks you probably have about this man of untold insight. He truly was one of the greatest inventors ever and he helped make America strong. His greatest flaws were trust and charity and because of those things, he spent his final years broke, misunderstood, and alone feeding and, he claimed, communicating with—the city's pigeons. He had broken the code between Magnetics and Gravity. He had a better understanding of Electromagnetic waves than any other including insights into cosmic emission, X-rays, how light becomes matter, how people affect reality, how to use the world as an electric conductor, and wave characteristics that developed what we feel as gravity. He understood more than any physicist that more energy is always input than is output from a body. This awareness allowed him to understand the major component of our universe that Dr. Keely, Einstein called Aether. To understand Nikola, we will have to examine some pretty unusual concepts that are only now being understood. We will even have to travel back thousands of years to see what was done before the massive mutations that occurred to humans 5 thousand years ago according to Haplotype DNA mutation scientists. While we are not certain what caused the massive mutations [1/2 of all major mutations of humans from the time they became human all happened at this time] we do know our brain began to shrink, we could now only use about 10% of our brain, we could no longer understand those not speaking our own language, our lifespan was shortened by about 80%, and both chimpanzee and Bonobo erupted as two of the mutations. What we will find is that many of the seemingly new discoveries are simply re-membered technologies. Mother Shipton, Dr. Cayce, Nostradamus, Tesla, and many others claimed that unseen "ancient teachers" showed them many of the marvels they reported on so we will look at that possibility as well. The following is a brief overview of those who affected or were affected by Nikola in their discoveries.

Inventor	Date	Description
John Reis	1861	He invented the telephone. The design praise would later be stolen by Alexander Bell.
James Clerk Maxwell	1831 - 1879	He developed a set of differential equations known as "Maxwell's Equations", which describe electromagnetic radiation and its interaction with matter. Tesla would use his work as a beginning of his discoveries.
Nathan Stubblefield	1892	He invented the ground battery, the first radio of sorts using low frequency magnetic transfer instead of high frequency RF signals and the first mobile phone using this same inductive transfer.
Thomas A. Edison	1847 - 1931	With 1,097 U.S. patents in his name, Thomas Edison is considered one of the most prolific inventors in history. His inventions included the stock ticker, phonograph, motion picture camera, the kinetoscope, and the incandescent lighting system. He became the antithesis of Tesla by greed, deception, and self adulation
John Keely	1871	Discovered the Aetheric force that controls the atomic constitution of matter and invented devices run on harmonic motion. The Aether would become the building block of all matter. Keely also began to understand how cognizant observation played a part in reality
Josiah Willard Gibbs	1878	He developed the theory of Chemical Thermodynamics introducing fundamental equations and relations to calculate multiphase equilibrium, the phase rule, and the free energy concept.
Wilhelm Konrad Roentgen	1895	This German physicist discovered a new kind of radiation working with the vacuum tube discharge. This radiation was called X-rays.
Max Plank	1899	He introduced the concept that light and all other kinds of electromagnetic radiation, which were considered as continuous trains of waves, actually consist of individual energy packages with well-defined amounts of energy quanta, proportional to its vibration frequency. All these concepts were initiated by Nikola Tesla.

Nikola Tesla	1856 - 1943	Nikola developed an alternating current system of generators, motors, and transmission lines. The radio, remote control vehicles, fluorescent lamps, fin-less turbines, X-Rays, VTOL airplanes, and resonant transfer of electricity through the earth were also his discoveries. He also worked with a whole lot more that may have been the beginnings of the understanding of vibrational matter, magnetic levitation, particle beam weapons, radar, remote powering of flying machines, and so much more.
Heinrich Hertz	1857 - 1894	He described the basic unit of frequency and built off Tesla's designs to improve electromagnetic transmissions and photoelectric effects. He also developed the spark-gap transmitter and dipole antenna.
Lee de Forest	1873- 1961	He made improvements on Tesla's vacuum tubes to produce the audio vacuum tube which amplified weak electric signals
Guglielmo Marconi	1874- 1937	He made improvements on Tesla's radio using 17 of Tesla's patents.
Einstein	1905	He introduced the Special Theory of Relativity, established the Law of Mass-Energy Equivalence, created the Brownian Theory of Motion, and formulated the Photon Theory of Light. He was friends with Tesla.
Wilson	1912	He described a cloud chamber that allowed the detection of protons and electrons. This also helped prove Tesla's details about Neutrinos/Cosmic rays.
Niels Bohr	1913	He proposed his "solar system" model of the atom and Quantum Mechanics.
Ed Leedskalinin	1935	He discovered a way to levitate heavy blocks using some type of magnetic device. It is not known if Tesla gained insight from his discovery.
John Hutchison	1979	He discovered a way to levitate of heavy objects, fuse of dissimilar materials such as metal and wood, while lacking any displacement, create the anomalous heating of metals without burning adjacent material, spontaneous fracturing of metals, change the crystalline structure and physical properties of metals, and even made a metal sample disappear using the high frequency, ultra-high voltage elements discovered by Tesla.

If you couldn't imagine life without your TV remote, thank

Nikola Tesla for making it possible. Tesla invented, predicted or contributed to development of hundreds of technologies that play big parts in our daily lives -- like the remote control, neon and fluorescent lights, the radio, computers, smartphones, laser beams, x-rays, flying, cosmic absorption, robotics and, of course, alternating current, the basis of our present-day electrical system. Certainly, he had a level of fame from some of his inventions, but I'm getting ahead of myself. First let me give you a fast and furious timeline before we get into how he made the world a better place and frontiered a new type of Physics to establish free energy.

Timeline of Nikola Tesla

1856: Born-Nikola was born to a Serbian family in the Austria-Hungarian land which is now called Croatia.

1863: Brother Dies- Dane at age 12 and traumatized 7-year old Nikola- begins seeing visions.

1874: Tesla Eludes Draft-Tesla roamed in the mountains for months to escape military draft.

1875: Tesla Attends College- Tesla made the highest grades possible but argued with professor, lost scholarship, and kicked out.

1882: Tesla envisioned AC current- From memory, he drew a diagram of the motor in the sand with a stick.

1882: Tesla Worked for Continental Edison- He submitted a plan for improving the Edison dynamos- was to be paid for the improvement, but not.

1883: Tesla fixed Lighting System- In France, Nikola helps the German Railway company. He was promised compensation, but none.

1883: Tesla's A.C. Motor-Tesla demonstrates his induction motor. <u>Investors could not see advantages and would not help</u>.

1884: Tesla in America- Now working for Edison, Nikola repaired an ocean liner's electric dynamos and other repairs and designed twenty-four different types of machines.

1885: Tesla Electric Light Company- While at Edison, investors got Tesla to improve the Arc Lamp and started Tesla Electric Light Company. <u>When finished the investors got the money and Tesla got nothing.</u>

1885: Tesla Quit Edison-Edison promises to pay him $50,000 to perfect his dynamo. <u>Nikola does and gets nothing so he quits.</u>

1896 Corona Discharge Device- Tesla was issued a patent for a corona discharge ozone generator using charged metal plates to act on ambient air for medical purposes.

1887: Tesla Electric Company- Alfred S. Brown, seeded Nikola to design stuff. First patent on the first month. company filed for its first patent by the end of the month.

1888: Tesla Sold A.C. Motor Patents-His A.C. Polyphase System patents were sold to George Westinghouse for $25,000 in cash, $50,000 in notes and a royalty of $2.50 per horsepower for each motor.

1890: Discovered Neon and Fluorescent lamps-With high frequencies, Tesla developed some of the first neon and fluorescent illumination.

1890: Discovered X-ray-He also took the first x-ray photographs of his foot and hand. <u>Roentgen would discover it a few years later and he got all the credit</u>. Tesla was not interested in this investigation after his assistant almost lost an arm and he almost lost an eye.

1890: Discovered Wireless Electric Power- transmitted energy through the air.

1890: Edison Revenge- Edison, mad about success of AC power- killed and elephant and even a convicted murderer showing how painful it was. <u>Later he would campaign for Marconi to be the creator of Radio even when he knew Tesla did it years before.</u>

1891: First Hydroelectric AC Power Station- Tesla established the Ames Power Plant and he became an American.

1891: Discovered Electrons- Tesla Wrote about the charges particle s in an atom well before J.J. Thomson confirmed them in another experiment. <u>Text books forgot about Tesla's discovery.</u>

1891: Invented the Tesla Coil -This high frequency high voltage emitter would be the basis for many new applications.

1893: Powered the Columbian Exposition- he also had massive display of electronic wizardry shown.

1893: Niagara Power Plant started- Often identified as the ending of the DC War, it was up and running by 1895 about the same time that Tesla's lab burned to the ground.

1894- Earthquake Machine- designed and tested and discredited until the 1980s in Canada.

1894- Sphincter Resonance discovered- Accidentally or on purpose, a specific vibrational tone was discovered that would loosen your bowels. This was discredited until re accomplished in the 1960s.

1896- Tesla Experiences Time Travel- Tesla is hit with a huge bolt of Electricity and he was saved by his quick-thinking assistant. Everything changed in time for a few seconds. This would spark interest in what was called Project Rainbow.

1898- Tesla Invented the Radio- He patented the radio and even demonstrated his radio using a remote-controlled boat. Marconi would not make his version for another 3 years. Tesla would not be acknowledged until after he died.

1898: Tesla's Robot Boat- Commanded by Radio signals, this completely submergible boat. Tesla demonstrated how to quickly win the Spanish American War, but no takers.

1899: Tesla Radio Receivers Hear Signals From Space- Tesla recorded rhythmic pulses from space. He only got criticized for claims of intelligence outside of the Earth.

1900: Current Around The Globe- In a lightning experiment he determined an electrical signal had traversed the Earth by resonating the signal to the natural resonance of the Earth.

1900: Tesla Ozone Company-Tesla began production of his corona discharge ozone generators to doctors who would have patients breath ozone bubbled through olive oil. He also made a o-zone/oil gel for doctors.

1901: Tesla Partners with J.P. Morgan- With $150,000 Tesla begins his World-Wide Wireless System known as

Wardenclyffe Tower. In 1903 massive lightning shot from the tower and streaks of light were everywhere according to witnesses. J.P. Morgan pulled funding and began an attempt to ruin Tesla. By 1912 the Tower is repossessed.

1910- Heat Pump Discovery- From many experiments Tesla patented several heat transfer manufacture of electricity for all types of applications including powering a boat.

1911- Wingless Air travel- Tesla had essentially discovered antigravity for air ships.

1915: Shared Nobel Prize – Both Edison and Tesla receive the Nobel Prize this year. The following year Tesla Declares Bankruptcy and the next year Tesla Receives Edison Medal.

1917: Tesla Describes Radar- He stated *"By standing electromagnetic waves use we may produce at will, from a sending station, an electrical effect in any particular region of the globe; with which we may determine the relative position or course of a moving object, such as a vessel at sea, the distance traversed by the same, or its speed."* He was not credited

1925: Katharine Johnson Died- Tesla's only love died. She married Tesla's longtime friend so they never married.

1927: Vertical take-off and land Air Travel Patent- Tesla patent 1,655,114 showed how to get more efficient aerial transportation. He is not credited.

1937: Tesla Hit By Taxicab-Thrown 35 to 40 feet, Tesla returned to his hotel and later stated, "It merely caused customary bruises and upset my digestion a bit." Three ribs were broken, but Tesla refused treatment and remained in his room for several months.

1937: Tesla Described his Death Ray-Realizing war was imminent, he described a charged particle beam weapon. Neither the U.S. War Department and European allies were willing to make the investment required to build the device.

1937: Project Rainbow- It was reported Tesla was contracted to head a group trying to make the USS Eldridge disappear from radar. He left the program by the end of 1942 as he considered it too dangerous. He was right.

January 8th, 1943: Maid Found Tesla Dead- Days before his sudden death he proclaimed he had solved the secret of Magnetism. Stacks and stacks of documents were seized by the FBI and in 1957 they were released, or, at least, most of them.

1980: John Hutchison Documented Tesla Experiments- With High frequency RF and Tesla Coil, John created unbelievable feats of Levitation, transmutation, invisibility and similar effects suggested by Tesla before his death.

This is just a tiny snapshot of somewhere between 300 and 500 patents, hundreds of technical papers, and thousands of experiments accomplished by the genius with an almost unbelievable photographic memory. He was mistreated, ignored, scoffed at, stolen from, misunderstood, and still he held onto a deep compassion for everyone.

9 yrs old   21 yrs old   29 yrs old   38 yrs old
50 yrs old   54 yrs old   59 yrs old   64 yrs old
70 yrs old   75 yrs old   79 yrs old   85 yrs old

The previous collage shows how he changed over the years. We'll not only track his character as he changes, and his genius, but also his compassion that best described him. Another inventor of his day was named John Keely. He actually established the name for the universal component of matter that is still used today [Aether] but he was defamed as he could not repeat many of his experiments and he had a limited understanding of magnetics. It would have been so very interesting if Nikola and John Keely could have worked together and that brings up another man named Ed Leedskalinin. Before we get to the details let's go back to the beginning.

John Keely [1837–1898] He discovered a "vaporic" or "Aetheric" and something he called "vibratory sympathy" which built "interatomic Aether". Interestingly, he and Tesla both we funded by John Astor before his fateful death on the Titanic. Below are his Negative Attractor and Iodicator, Compound Disintegrator, Sympatheric Negative Attractor, and his Vibratory Globe and Accelerator. Unfortunately he was not using high frequency, ultra-high voltage AC power which certainly would haveallow his work to be expanded more reaosnably.

Ed Leedskalinin [1887 - 1951] Like Nikola, Ed was given instruction on designing an antigravity machine by some ancient invisible teacher. He said he had solved how the Egyptians raised the massive stones on the pyramid. This same knowledge allowed him to raise hundred-ton stones in the air without help. The only thing left of his machines are shown below. Massive numbers of magnets, must have built up some strange force. Tesla could have used this effect in his designs.

Thomas Edison [1847 –1931] This guy didn't really invent anything new but was in the habit of stealing other people's works and patenting them. He did this for the light bulb, phonograph, motion picture machine and many others. In 1890, Le Prince, the inventor of the motion picture machine, was taking a trip to patent his invention in England. He got on a train on September 13, 1890, and was never seen again.

The family continued with the patent quest. Unfortunately, in 1892, while Le Prince's son was testifying in a patent trial against Edison, the son was mysteriously shot to death by an unknown assailant. This murder was never solved. Along this same line, filmmaker Georges Melies's masterpiece "*A Trip to the Moon*" was spreading like wildfire throughout London. Edison obtained an illegal copy from a shady theater owner, made numerous copies, took them to America, showed the pirated film across America, and reaped huge amounts of money. When Melies arrived in America to show the film, everybody had already seen it. This loss directly led to his bankruptcy. Below are some of his famous frauds. Remember this as we investigate Nikola. I suppose one could say he got off lucky.

Off course Edison was not the only one who would reap the benefits of ripping off the sometime naive Nikola. J.P. Morgan, Edison, Westinghouse, and many others hated that Nikola wanted to help humanity rather than fill his pockets. Not only did it make them look bad, but also it reduced their profits. They were willing to sacrifice our future at all costs. Luckily, he had true influential friends like John Astor IV, Einstein, Mark Twain who did not abandon him.

1856 Strange Beginnings

Strangely, Nikola Tesla was born on July 9 and 10, 1856, during a thunderstorm, in what is now Smiljan, Croatia. To make his entrance memorable, he began leaving his mom before midnight on the 9th and did not fully separate from his mom until after midnight which became the next day. That would only be the beginning of a strange childhood. A copy of his birth Certificate is shown below.

He was one of five children which included siblings Dane, Angelina, Milka and Marica, in the family. Tesla's interest in electrical invention was spurred by his mother, Djuka Mandic, who invented small household appliances in her spare time kind of like MacGyver on that TV show making outlandish things from a gum wrapper, I suppose, while Nikola watched. Tesla's father, Milutin Tesla, was a Serbian orthodox priest and a writer, and he pushed for his son to join the priesthood. In fact, they live behind the church as shown below. By the way, don't worry about the Serbia Croatia thing, the country was Serbia and later became Croatia.

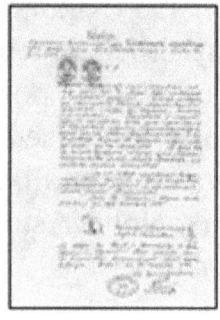

While Nikola did not go into the priesthood, he was a God-fearing man and stated the following:

The gift of mental power comes from God, Divine Being, and if we Concentrate our minds on that truth, we become in tune with his great power.

When he was 7, in 1863, Tesla's brother Dane was killed in a riding accident. The shock of the loss sent Nikola over the edge and he began to see visions. This would be the first signs of his lifelong mental illness or genius. Anything he had done or seen became part of the visions. Possibly, he had been infected or gifted with Hyperthymesia or Highly Superior Autobiographical Memory (HSAM). This rare condition only has about a dozen "infected" people in the world today. Marilu Henner [TV star] is one and Nikola might very well have had the same condition. Today, we know that this condition allows the person not only to have a photographic memory, but also, the person never forgets anything.

The images above are Marilu in the long running TV series "Taxi" circa 1979 and Nikola about the time he started getting these visions. Their similarity is astounding [memory-wise I mean]. Anyway! Nikola's interests drifted from the Greek Orthodoxia priest life and went towards the sciences. After studying at the Realschule, Karlstadt; the Polytechnic Institute in Graz, Austria; and the University of Prague during the 1870s, Tesla moved to Budapest, where for a time he worked at the Central Telephone Exchange. He tells us it was while in Budapest that the idea for the induction motor first came to Tesla. The image below left shows Tesla during his university years. Yes, it seems he was related to Jimmy Stewart shown playing the accordion and just staring. I checked the records and unless Serbia and Ireland are near each other, the similarity is unfounded.

After several years of trying to gain interest in his induction motor invention, at age 28, Tesla decided to leave Europe for America to work for Thomas Edison in 1884.

Visions

Nikola was different. Instead of drawing up plans, erasing and redrawing, and repeating the process to obtain a design, he simply did it all in his head. Let's just read what he said about his process.

Before Inventing came to his Life

During my boyhood I had suffered from a peculiar affliction due to the appearance of images, which were often accompanied by strong flashes of light. When a word was spoken, the image of the object designated would present itself so vividly to my vision that I could not tell whether what I saw was real or not. Even though I reached out and passed my hand through it, the image would remain fixed in space. In trying to free myself from these tormenting appearances, I tried to concentrate my thoughts on some peaceful, quieting scene I had witnessed. This would give me momentary relief; but when I had done it two or three times the remedy would begin to lose its force. Then I began to take mental excursions beyond the small world of my actual knowledge. Day and night, in imagination, I went on journeys — saw new places, cities, countries, and all the time I tried hard to make these imaginary things very sharp and clear in my mind. I imagined myself living in countries I had never seen, and *I made imaginary friends*, who were very dear to me and really seemed alive.

Invention Saved Him

This I did constantly until I was seventeen, when my thoughts turned seriously to invention. Then to my delight, I found I could visualize with the greatest facility. I needed no models, drawings, or experiments. I could picture them all in my mind. By that faculty of visualizing, which I learned in my boyish efforts to rid myself of annoying images, I have evolved what is, I believe, a new method of materializing inventive ideas and conceptions. It is a method which may be of great usefulness to any imaginative man, whether he is an inventor, businessman or artist. Some people, the moment they have a device to construct or any piece of work to perform, rush at it without adequate preparation, and immediately become engrossed in details, instead of the central idea. They may get results, but they sacrifice quality. Here in brief, is my own method: after experiencing a desire to invent a particular thing, I may go on for months or years with the idea in the back of my head. Whenever I feel like it, I roam around in my imagination and think about the problem without any deliberate concentration. This is a period of incubation.

Experimentation

Then follows a period of direct effort. I choose carefully the possible solutions of the problem I am considering, and gradually center my mind on a narrowed field of investigation. Now, when I am deliberately thinking of the problem in its specific features, I may begin to feel that I am going to get the solution. And the wonderful thing is, that if I do feel this way, then I know I have really solved the problem and shall get what I am after. The feeling is as convincing to me as though I already had solved it. I have come to the conclusion that at this stage the actual solution is in my mind subconsciously though it may be a long time before I am aware of it consciously. Before I put a sketch on paper, the whole idea is worked out mentally. In my mind I change the construction, make improvements, and even operate the device. Without ever having drawn a sketch I can give the measurements of all parts to workmen, and when completed all these parts will fit, just as certainly as though I had made the actual drawings. It is immaterial to me whether I run my machine in my mind or test it in my shop. The inventions I have conceived in this way have always worked. In thirty years there has not been a single exception. My first electric motor, the vacuum tube wireless light, my turbine engine and many other devices have all been developed in exactly this way."

I told you he had perfect recall and a photographic memory. Pretty neat. That was good when designing hundreds of inventions at the same time, but what about women?

1884 Tesla' Women

The one girl he loved throughout his life is shown above and to the right following; Katharine McMahon, who married his longtime friend Robert Underwood Johnson instead of him. They stayed very close until her death in 1937.

Besides, noticing women, the first thing Nikola seemed to do in his early 30s was grow a mustache so he would not look like Jimmy Stewart who wasn't even born yet, as shown right. While it appears, he was a lady's man in the following image to left, he had issues with many women and never married, but he did enjoy sitting on the beach with a girl and relax in fancy swim-ware. Possibly the girl in the picture was his beloved Katharine, shown again to the right. He could not stand to see pearls, or touch hair, so there were limits to his level of intimacy.

All that being said Tesla was a lady's man who dated high society women. Those most known were Flora Dodge and Sarah Bernhart shown next left and right.

While he had interaction with women, it was not often and he always went back to his true love Katharine McMahon Johnson and she asked her husband to take care of Nikola on her deathbed. Speaking of death, I have to get back to Edison.

1884 Tesla and Edison

When Tesla went to work for Edison in 1884, he installed and repaired incandescent and arc lamps, reassembled D.C. generators and designed twenty-four different types of machines including a massive Edison dynamo from the Pearl St. New York Station. [See below] Working for Edison was an eye opener for Nikola. Edison was having horrible problems with his inefficient DC motor and offered to pay Tesla $50 thousand dollars if he could just find the problem. Tesla quickly improved the motor with great ease and allowed Edison to reap substantial profits.

When Edison was to pay him, Edison laughed and told Tesla that he *"failed to understand the American sense of humor."* Even with Edison offering to increase his pay from $18 a week to all of $25. Tesla, quit and started his own company.

1885 Tesla Electric Company

In 1885, Nikola had received funding for the Tesla Electric Light Company and was tasked by his investors to develop improved arc lighting. Samples of his new Arc lamps are shown next but being taken advantage again was coming up again.

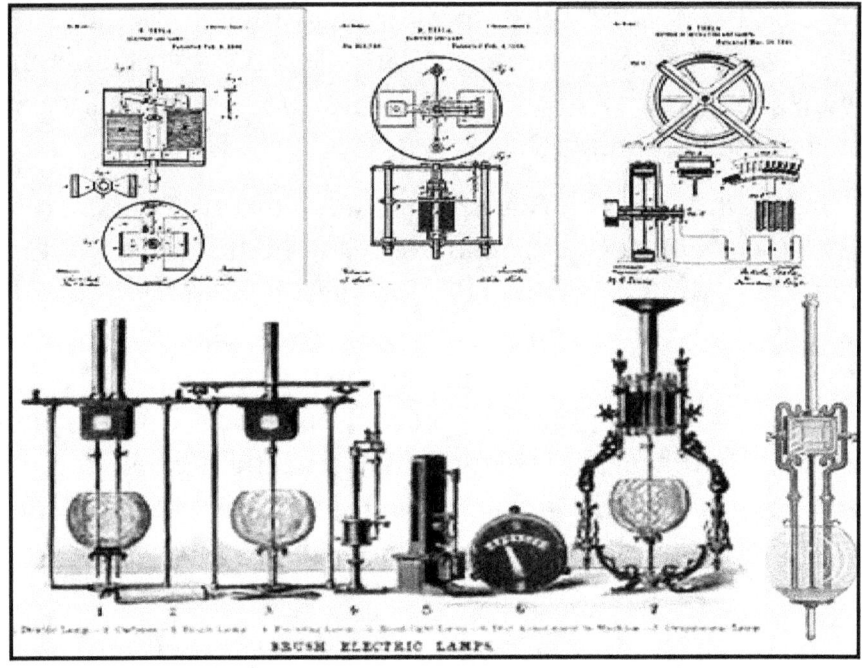

After successfully making the Arc lamps, Tesla was forced out of the venture and for a time had to work as a manual laborer in order to survive. Ditch digging was not his thing and his luck changed in 1887.

1887 Tesla's New Electric Company

AC Electricity was not only worked on by Nikola, but he may have been the only one who was getting inside information from invisible people and had the ability to store details, draw models in their head, or remember exactly what they had been working on in their head for a long time. The idea was actually presented in Europe by Guillaume Duchenne [circa 1870] but there were substantial issues with these crude beginnings.

Tesla, on his own developed a model of an AC generator with no communitator around 1880 and he had an awful time trying to get interest until 1887, in fact, Edison tried to quiet his rantings as not good for business every chance he got. In 1887 Tesla was able to find interest in his AC electrical system and funding for his new Tesla Electric Company. Investors had seen his rotating magnetic field cause a metal egg to begin spinning on a plate and knew there was a genius in front of them. [See the egg rollers device below.]

Setting straight to work, by the end of the year, Tesla had successfully filed several patents for AC-based inventions. One version of his AC motor is shown with some of tesla's lightbulbs next to it.

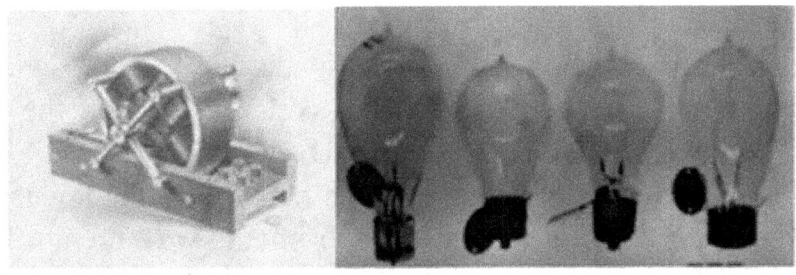

The diagram below is part of his patent on this first unit.

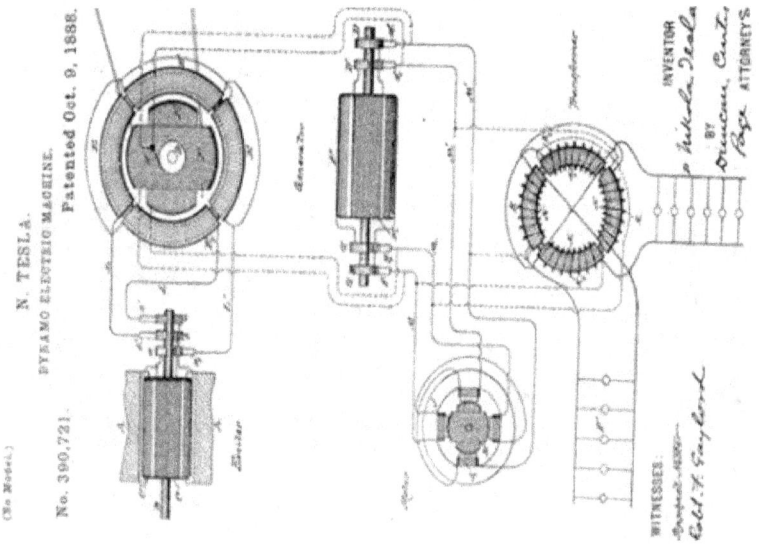

Almost immediately Edison launched a campaign against AC, claiming it was dangerous and could kill people. Tesla countered by publicly subjecting himself to 250,000-volt shocks to demonstrate AC's safety. We'll get to that in a minute. Right now, Tesla is taking pictures of his own guts.

1890 Tesla X-Rays

Tesla simply called it "a very special radiation" when he invented something to look at skeletons under the skin. While he was working his "carbon-button" lamp. Well before 1895 He produced pictures he called "shadowgraphs" and had performed numerous experiments with them up until the fire that destroyed his lab also destroyed his X-ray equipment. Upon learning of Röntgen's discovery, Tesla wrote him and sent some of the pictures recovered from the fire. Röntgen replied and asked Tesla how he produced them. Tesla's foot and hand are shown below.

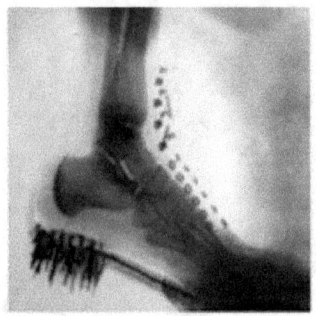 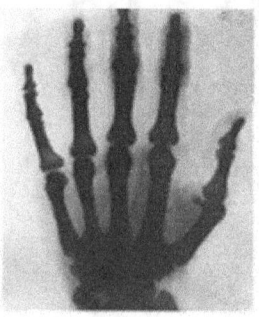 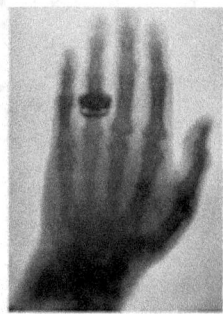

Next, we see Tesla's head followed by the second inventor of X-ray, Röntgen.

The following is another Nikola X-Ray published in the Electronics Review magazine in the 1890s.

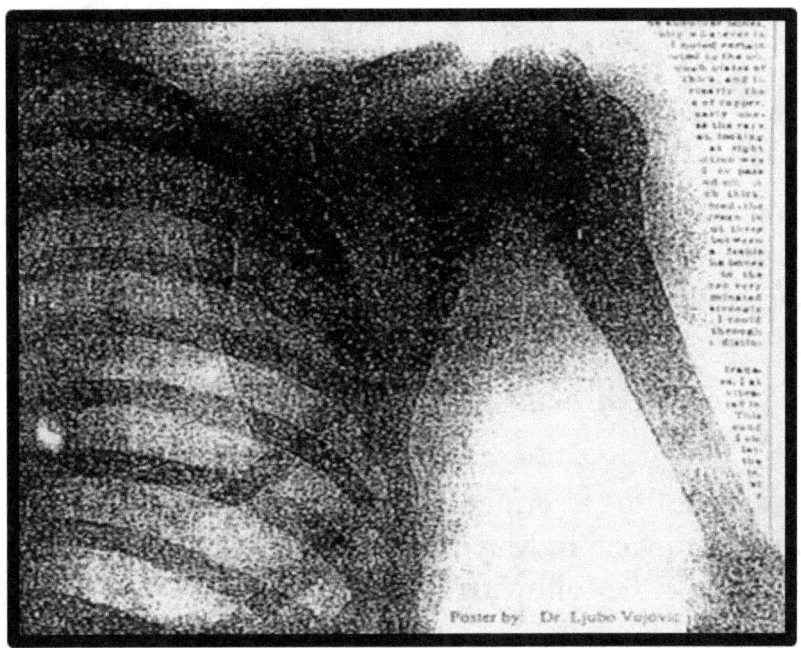

Poster by Dr. Ljubo Vujovic

Images of Niklola's improved Lenard/Cathode Ray tubes for production of X-Ray are shown next. This 1998 patent would further X Ray developments, increase imaging range, and make clearer images. Nikola indicated he could porject X-Rays out over 40 feet.

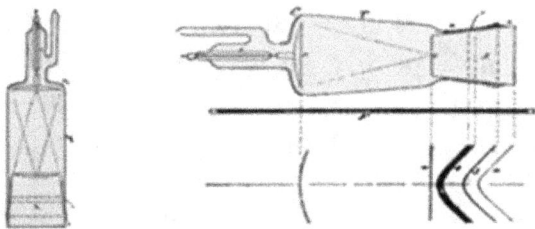

Images following show various setups for X-Ray photography Tesla used, but soon he halted all his experiments.

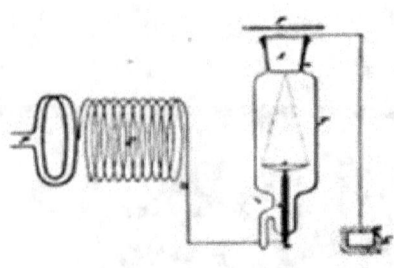

Edison, had forgotten how nasty he had been from 1885 until 1891 and sent this message to Tesla on 18 March 1896 when Tesla wrote about an essay that was to be written:

"My dear Tesla, Many thanks for your letter. I hope you are progressing and will give us something that will beat Roentgen."

In response to a critical essay to be published in the *Electrical Review* in May of 1896, Edison sent a note to the writer saying the following:

"Tesla is of a nervous temperament and it will greatly grieve him and interfere with his work. While Tesla gives vent to his sanguine expectations when he should not do so, it must not be forgotten by [the article author] Mr. Moore that Tesla is an experimenter of the highest type and may produce in time all that he says he can."

Nikola sent back a somewhat nasty letter indicating Edison had forgotten how poorly he was treated and that he had no intentions of continuing X-Rays as he had almost lost one of his eyes and his assistant had almost lost and arm.

While the letters from 1896 sound like Edison had deep respect to the genius of Tesla, Tesla 's remarks were about how Edison tried to destroy him in what has been called the *War of Currents*.

1890-War of Currents

Well before the 1896 letter, Edison was in bitter attack mode to eliminate people's desire for AC current product. A war quickly started as Nikola gained moderate success. Edison know that if people got comfortable with the more efficient power source, he would lose many customers. AC systems were easier to transport, to combine, to control, and had less resistance being carried long distances on power lines. Edison though if he could convince his newspaper friends the electricity was too dangerous he would have a chance. In 1890 his diabolical mind found a prisoner to kill in a horrible way. The device and William Kemmler are shown next. It would take a while for him to die.

In 1903 Edison cooked a live Elephant that had killed someone. Smoke spewed out of the Elephant before he finally fell over, as shown in the next 2 images.

The Prisoner execution event was just as gruesome and drawn out. George Westinghouse was quoted to have said, *"They would have done better using an axe."* And so, the first execution by electric chair took place, just to prove Tesla wrong (and preserve Edison's financial stakes).

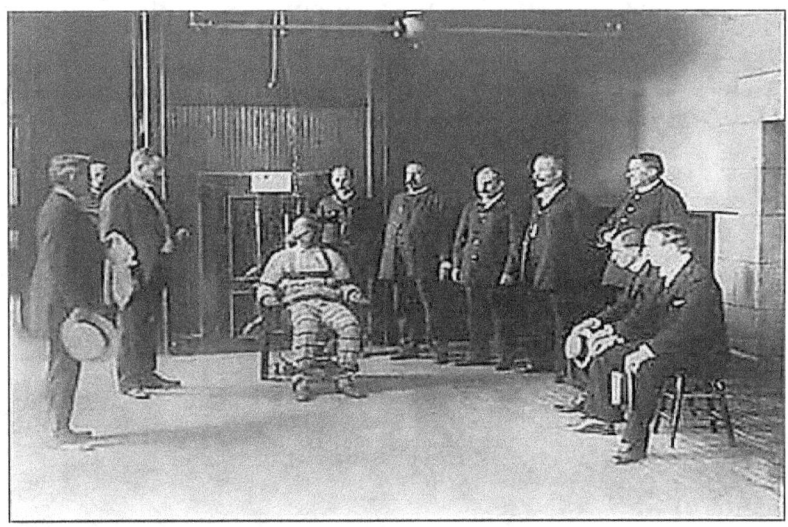

Tesla simply ignored Edison and made something he called a Tesla Coil.

1890 Electricity Thru Air

While in some ways the Tesla Coil was not the most monetarily successful of Tesla's invention, it is certainly the invention he is most famous for and some of the most amazing feats were established with the Tesla Coil as a major part.   The Tesla coil was originally developed to power Tesla's new wireless lighting systems, but later became the basis of the ill-fated World-Wide Wireless System, otherwise known as Wardenclyffe. The image below shows Tesla demonstrating his Tesla Coil experiments before the Royal Society. Later we will see it used for other important discoveries.

Besides great potential for all types of discoveries, the Tesla coil give Tesla instant fame as it seemed that he could make electricity with his hands as shown next.

Typical emissions of high voltage, high frequency electricity from Tesla Coils are shown below.

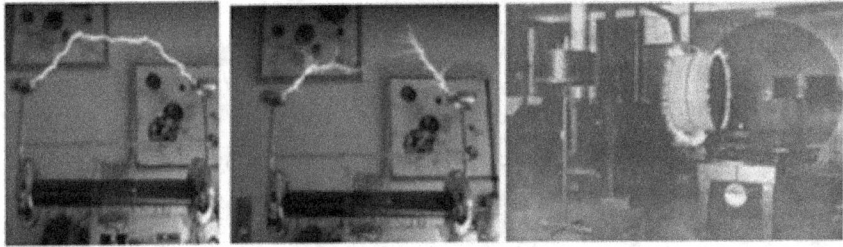

At one time he passed 250 thousand volts through his body and audiences raved. The following drawing commemorated that event of shock and amazement.

According to a newspaper article, *The observer holds a loop of bare wire in his hands. The currents induced in the loop by means of the —resonating— coil over which it is held, traverse the body of the observer, and at the same time, as they pass between his bare hands, they bring two or three lamps held there to bright incandescence.* [See image right]

Strange as it may seem, these currents, of a voltage one or two hundred times as high as that employed in electrocution, do not inconvenience the experimenter in the slightest. The extremely high tension of the currents which Mr. Clemens is seen receiving prevents them from doing any harm to him. [See image left]

The Tesla Coil had become a thing of awe inspiring intrigue. It had made high frequency High voltage electricity, so Tesla invented Fluorescent Lights.

1892 Florescence

Now for the first commercial use of the high frequency, high voltage discharge. High voltage could be used to excite materials to establish a glowing plasma. The images below show some of his many displays of through air transmission of energy to excite his fluorescent bulbs. These first devices were called phosphorescent lamps. Now we use them everywhere.

  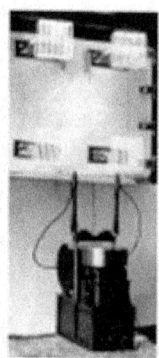

The first time we read about electrical plasma discharges is from extremely ancient texts describing the planet Venus as it began to glow and show off a fiery tail as it was being destroyed 11 thousand years ago when the Planet Venus and Earth came close enough to each other so that their electrical differences were neutralized at the cost of Venus being destroyed. Possibly, Nikola knew about this plasma or he may have been told about it from his invisible teachers, or he just thought it up on his own, but here are some details about the massive plasma strings similar to the capture plasma in a fluorescent bulb that was formed between planets.

Interestingly, the plasma is still somewhat visible. Plasma is simple ions in air [or space] that have been excited by high voltages once excited the ions appear to be conductor of electricity, sort of like an invisible wire. As electricity passes, the ions heat up and can glow. While that is interesting and allows us to have 4-foot-long lights for our kitchen, we are going to talk about a plasma string that is still 40 million kilometers long and was once 45 million kilometers to connect Earth and Venus.

Huge Plasma Tail

As we very briefly describe this critical time for the planet Venus, let me say that while most planets have a plasma covering that is mostly invisible to our eyes, they are typically tiny. Venus, on the other hand, has a measured plasma sheath that is two or three times the planet's diameter – [say about 20,000 miles wide and almost the 26 million miles long that is the distance between Venus and our planet]. So, the Venusian tail is approximately a thousand times as long as it is broad at its thickest point. It would look like the planet had a wavy tail if the plasma was excited to a level that allowed it to become visible. The SOHO satellite recorded the massive plasma strings that almost travel all the way to Earth as if at one time, Earth somehow electrified Venus. [See following image]

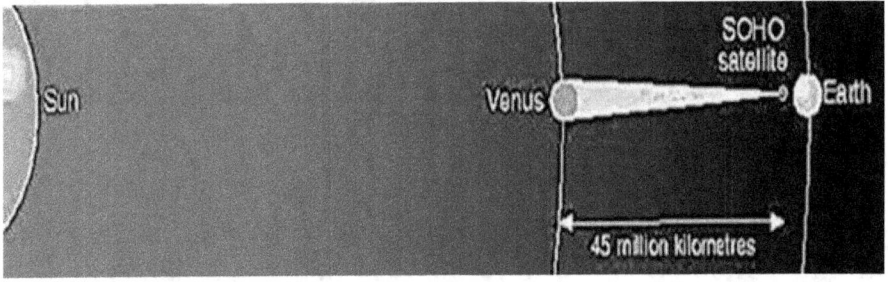

The reason I am providing this background is to potentially show where Tesla may have gotten some of his ideas. Two conclusions arise; the electrification was done by divine providence, or by shear bad-luck. What I mean by that is that each planet is a massive charging station of electrical potential. As it spins it builds up this electric field. This doesn't matter because there is no place to discharge the electricity so it can generate NO current. The problem is that Venus got too close to Earth and possibly its position got synchronized with Earth's long enough for the plasma strings to form and once there was an electric path, a massive lightning bolt would be discharged. It is believed the Venusian moon was hit by the lightning bolt/plasma and shattered. This sent out millions of pieces that hit Venus to slow its rotation and the rest is well known. Please notice the massive crack in Venus showing it almost split apart during its destruction. Possibly, Tesla's unseen teachers described all this to him showing the power of electrical plasma.

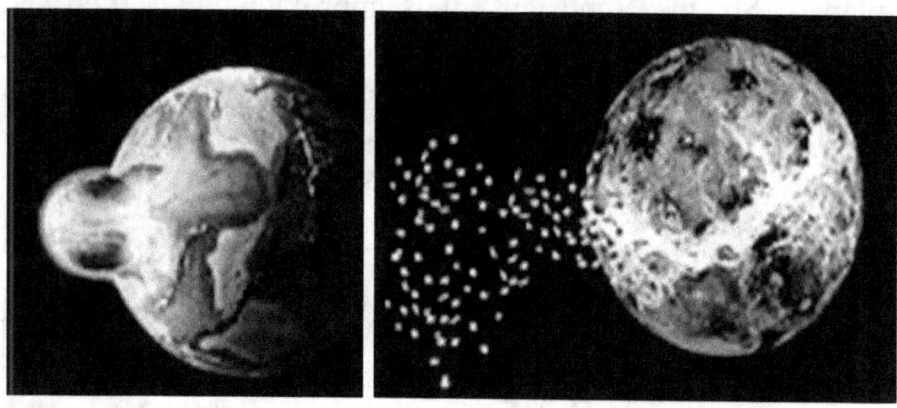

By the way, we use electrical plasmas all the time in florescent lights and all types of things. Eleven thousand years ago, electrical plasma may have killed Venus. You might wonder why we time the destruction of Venus and its plasma tail to 11 thousand years ago, so here are some details.

Half a million craters- on the East Coast of the United States, called the Carolina Bays came from somewhere close in our Solar System. They are date to about 11-thousands years ago.

Something is Making Sulfur-Dioxide-All I can say is three things- Sulfur-Dioxide, Carbon-monoxide, and Ozone. Using data from the Russian Veruna space missions and also the US Pioneer and Magellan probes, researchers studying the high concentration of water droplets in the Venusian clouds found hydrogen sulfide and sulfur dioxide. These two gases react with each other, so they should not be seen in the same place unless something **IS** producing them. This also suggests Venus was not destroyed long ago.

Something using Carbon Monoxide-Despite solar radiation and lightning, the atmosphere contains hardly any carbon monoxide. This suggests something is removing the gas and its destruction was not long ago.

Ozone Indicates Current Life -One belief is that bugs living in the Venusian clouds could be combining Sulfur dioxide with carbon monoxide and possibly hydrogen sulfide or carbonyl sulfide in a metabolism similar to that of some early Earth bugs. The "Venus Express" [2008 to 2012] found Ozone in 2011. This doesn't make sense in that it has only previously been detected in the atmospheres of Earth and Mars. On Earth, it is of fundamental importance to life because it absorbs much of the Sun's harmful ultraviolet rays. Not only that, it is thought to have been generated by life itself in the first place and suggests Venus destruction was

43

not long ago a while Venus was dying, a massive plasma discharge was seen by everyone on Earth.

Greek legend, *"A blazing star almost destroyed the world with fire before it became Venus."*

The Inca legends tell the story. The Inca called Venus the "*Wavy haired planet*";

The Aztecs called Venus *"the Star that smoked" and said that it once passed by the world blazing and killing many people.*

Ute Indians tell us the same thing in their verbal history. *"The sun was slivered into a thousand fragments, which fell to Earth causing a general fire.*

Blackfoot had to say. According to their traditions, *"The morning star [Venus] put on a scarlet cloak [sounds like it turned red.] and appeared before a woman on Earth that he loved. She went into the sky with him, but was warned never to look back. She did, of course, and was ordered to return to Earth."*

Babylonian Texts- *"To the queen of the heavens Inanna [Venus], to her who filled the sky with her pure blaze. The laminations are as bright as the sun. Who initiated the flood-storm? You roared in the heavens and Earth. You smote the flesh of the people." My hair will whirl in heaven for you." You flash like lightning over the highlands. You throw firebrands across the Earth.* As the Pleistocene Extinction caused a huge worldwide flood we can believe it happened before the extinction.

Sumerian Tests Ishtar, but it is the same. *"To the pure flame that fills the heaven, who shines like the sun 'Ishtar"*

44

[Venus]—"I ran battle down like flames in the fighting. I make heaven and Earth shake. I trample the Earth. I destroy what remains of the inhabited world".

Phoenician texts describe the event, but this time the goddess is Astarte, the Phoenician version of Ishtar. *"See, Astarte" [Venus], she descends into a pool as a fiery falling star"*

Greek Tests- *"Her [Venus's] anger grew so terrible that she transformed herself, grew smaller and black. On a blind rampage she was killing everything and everyone in sight. Her hair is wild, her eyes red. The world trembles and cracks under her tread. Her dark hair flies in the sky sweeping away the sun and stars."*

Chinese writers *said the same thing, "There was a time when a planet [Venus] approached close to the Earth, causing great showers of stones."* [Not too many of the planets could have come close to earth. The moon of Venus is my guess.] *Venus was depicted as a dangerous "fire spitting planet"*

Samoan History-*Venus was depicted as a dangerous, "fire spitting, planet".*

All these sightings tell us the Venus destruction happened during a time when people could pass down the sightings of the visible plasma streamers from the planet.

Tesla's Fluorescence

We can believe Tesla had heard or read about, at least, some of these events and with his understanding of magnetics, gas discharge made his discernment fairly simple. Tesla's fluorescent lamps or fluorescent tubes were low pressure mercury-vapor gas-discharge lamps that used fluorescence or something we call electrical plasma to produce visible light. An electric current in the gas excites mercury vapor which produced short-wave ultraviolet light that then caused a phosphor coating on the inside of the bulb to glow just like Venus had glowed. A fluorescent lamp converts electrical energy into useful light much more efficiently than incandescent lamps. The luminous efficacy of a fluorescent light bulb can exceed 100 lumens per watt, several times the efficacy of an incandescent bulb with comparable light output so this was a valuable commodity. Some of Tesla equipment for this lighting is shown next.

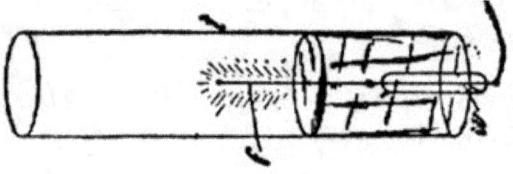

A modification of his Tesla Tube would excite the gas. The device is shown next.

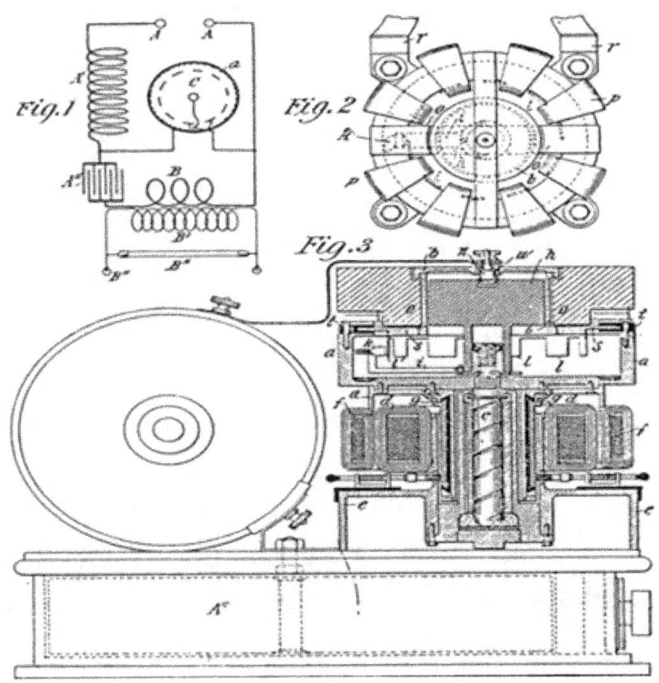

Once Tesla got enough light and sent plasma though the air, he started working of sending messages through the air.

1892 First Radio

The images below are early radio transmitters and receivers invented by Tesla. The little plug in oscillator shown in the image to the right allow the radio to be "tuned" to a desired channel. Tesla also boasted that Marconi's later version used 17 of his patents to make his version of "radio transceiver" work so there cannot be question concerning the limited work of Marconi in this area. Even before these designs, Tesla had designed and demonstrated the radio concept by sending RF messages through the air to light lamps as shown in the middle image. Later he would send messages to operate a "remote controlled Robotic boat". With great sadness we find almost no mention of his achievements and some stupid reference of Marconi inventing the Radio, which is a complete LIE.

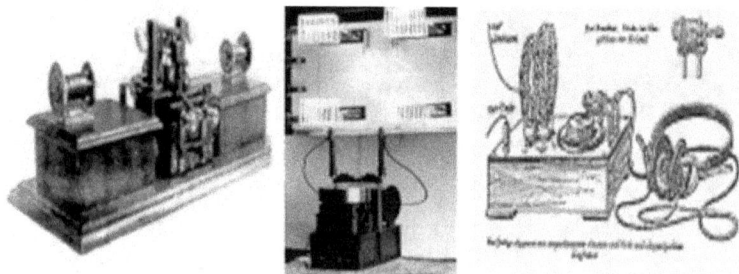

Besides all of this, Tesla's system of wireless transmission thru the earth was actually exposed in his lectures before the

Franklin Institute and Electric Light Association in February and March, 1893 as pictorially shown from Patent drawings.

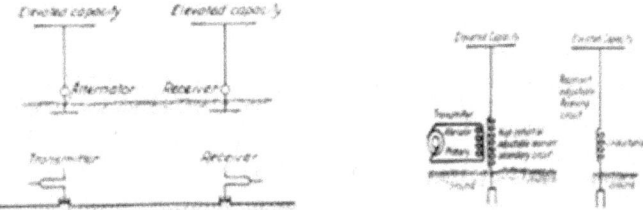

Addititional details first show Tesla's patent [left following] followed by Marconi's patent that was virtually identical [right]. The Supreme Court would overturn Marconi's patent after Tesla's death in 1943, but no one seems to care.

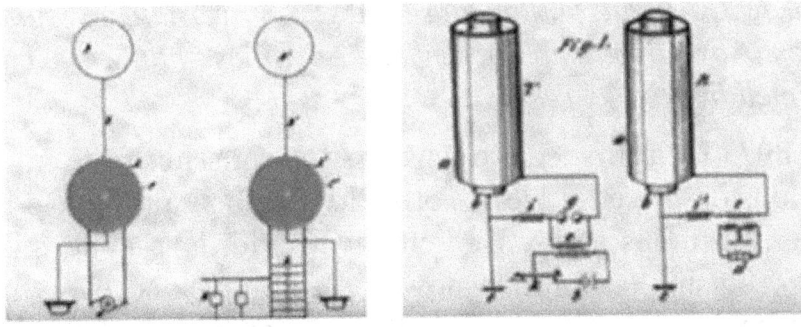

Both patents tuned an antenna to one frequency to differetiate signal transmission, however, doubts arose among Tesla's investors about the plausibility of Tesla's system so his rival, Guglielmo Marconi, with the financial support of two traitors to America, Andrew Carnegie and Thomas Edison, continued to make great advances with his own radio technologies for Italy.

Tesla is now credited with inventing the modern radio; since the Court overturned Guglielmo Marconi's patent in 1943 in favor of Nikola Tesla's much earlier patents.

One of the 17 Nikola patents used by Marconi, the Tesla coil, invented in 1891, is still used in radio and television sets and other electronic equipment. Let me just say this once more. Guglielmo Marconi was initially credited, and most STILL believe him to be the inventor of radio to this day. Tesla's initial demonstration of the "Radio" concept was in 1893 during a presentation before The National Electric Light Association. In 1897 Tesla applied for two patents US 645576, and US 649621 specifically for Radio transmission, but *in 1904, the U.S. Patent Office [Temporarily] reversed its decision, awarding Marconi a patent for the invention of radio, because of financial backers Thomas Edison and Andrew Carnegie and because the U.S. government would be able to avoid having to pay the royalties that were being claimed by Nikola Tesla.*

In 1897, Tesla first sent a wireless transmission from his lab at Houston Street in New York City to a boat on the Hudson River, 25 miles away. He actually would have done this in 1895, but his lab burned down. Tesla invented everything we associate with radio -- antennas, tuners and the like, so all textbooks you read talking about Macroni's INVENTION are absolutely LYING. Following is a brief description of Tesla's Radio transceivers and his earlier work. [I apologize for the ranting.]

"The chief object of employing <u>very short waves</u> is to provide an increased <u>number of channels</u> required to satisfy the ever-growing demand <u>for wireless appliances</u>. But this is only because the transmitting and receiving apparatus, as generally employed, is ill-conceived and not well adapted for selection. The transmitter generates several systems of waves, all of which, except one, are useless. As a

50

*consequence, only <u>an infinitesimal amount of energy reaches the receiver</u> and dependence is placed on extreme amplification, which can be easily affected by the use of the so-called <u>three-electrode tubes</u>. This invention has been credited to others, but as a matter of fact, **<u>it was brought out by me in 1892</u>**, the principle being described and illustrated in my lecture before the Franklin Institute and National Electric Light Association. In my original device I put around the incandescent filament a conducting member, which I called a "sieve." This device is connected to a wire leading outside of the bulb and serves to modify the stream of particles projected from the filament according to the charge imparted to it. In this manner a new kind of detector, rectifier and amplifier was provided. <u>Many forms of tubes on this principle were constructed by me</u> and various interesting effects obtained by their means <u>shown to visitors in my laboratory from 1893 to 1899</u>, when I undertook the <u>erection of an experimental world-system wireless plant at Colorado Springs</u>.*

Without the evil of Andrew Carnegie [who had become one of the wealthiest people in America because of his part in the Civil War] and Thomas Edison [who had gained wealth by stealing inventions of others], Nikola Tesla would never have had his patents taken away and our textbooks would not all be wrong.

Before all this happened, Tesla and Westinghouse had put in a massive AC generating plant and were invited to power the 1892/3 World's Fair in Chicago.

1893 World's Fair

In 1892 George Westinghouse won the contract to power the Columbian Exposition. The Westinghouse company, with Tesla's guidance, built a power system for the exposition that produced three times more energy than was being utilized by the entire remainder of Chicago. Tesla had a large display including incandescent and phosphorescent lighting (the precursor to fluorescent lamps) powered without wires by high-frequency fields and the Egg of Columbus that was used to restart his first company. See below right. The success of the Tesla Polyphase System installed at the exposition ensured Westinghouse be successful. The Westinghouse A.C. switchboard was used to power the fairgrounds and Tesla's display at the Columbian Exposition is shown below left.

Because of the massive success at the fair, on October 24th, 1893, the Niagara Power Plant Contract Awarded to Nikola and Westinghouse.

These were his hand on face years

The images below show some of his favorite stances. The first is the 2 closed-finger stance. Next, we see his three-finger stance.

That is followed by the 2 open-finger stance. Finally, the three Open-finger stance shows his flexibility in how he wanted to be photographed.

He also tried his best to get people to shoot his "good" side and he began looking at the sun more intently.

1893 Power Plant

The Ames Hydroelectric Generating Plant, constructed in 1890 near Ophir, Colorado, was the world's first commercial system to produce and transmit AC electricity for industrial use, the first electrical plant to use Tesla's induction motors, and one of the first AC hydro-electric plants ever constructed. The Niagara Fall Power plant would be the 2nd major hydroelectric power plant using Nikola's inventions. The new A.C. power system enjoyed a flawless inauguration, transmitting electricity to Buffalo, New York 22 miles away. It came first to the Buffalo Railway Company - 1,000 horsepower, switched into the company's powerhouses at exactly midnight with a signaling of the event to the city by the firing of cannons, the blowing of steam whistles and the ringing of bells. The machines are shown below.

Next, we see the infrastructure for the plant.

During all of this success, Nikola set up a laboratory in New York and looked to expand his awareness. In a nutshell, Tesla determined that many strange things could be produced with high-energy vibrations of electricity. He was even being told how to accomplish some of his amazing experiments by unseen "helpers". If you go to the *"Edison Tech Center"* internet site, here is what it says about the Niagara Plant and Tesla. Please notice, even today, Nikola's name is continuously removed from all things Edison.

1895 - Westinghouse builds the power system for the Adams Power Station at Niagara Falls. Benjamin Garver Lamme is the principal engineer of the operation. General Electric builds the 25-mile power transmission system from the Niagara power house to Buffalo, NY which is made operational in 1896.

I'm trying to not get angry and simply believe these people are misguided. Anyway! The Niagara Plant would soon be powering New York City and then something strange happened.

1895 Tesla's Time Travel Experience

All this first stuff seems great, but one might believe that even a normal person might be able to achieve these remarkable things, but when it comes to time travel, we start seeing the real genius and daring of Nikola. Let me say there are questions about this whole incident, but we can be sure of one thing. Nikola NEVER forgot it for 3 reasons. He almost died, he could never forget anything, and he saw something strange.

Time Travel- I suppose the first we hear about Tesla working on time dilation and travel in March 1895. A reporter for the *"New York Herald"* wrote on March 13th that he came across the inventor in a small cafe, looking shaken after being hit by 3.5 million volts, *"I am afraid,"* said Tesla, *"that you won't find me a pleasant companion tonight. The fact is I was almost killed today. The spark jumped three feet through the air and struck me here on the right shoulder. If my assistant had not turned off the current instantly in might have been the end of me."* Tesla, on contact with the resonating electromagnetic charge, <u>found himself outside his time-frame reference</u>. He reported that **he could see the immediate past – present and future, all at once**. But he was paralyzed within the electromagnetic field, unable to help himself. His assistant, by turning off the current, released Tesla before any permanent damage was done.

While the immediate experiment had been to get electrical power from the air, the effect was somewhat different. Below is a drawing of the general setup used with substantially different results than initially planned.

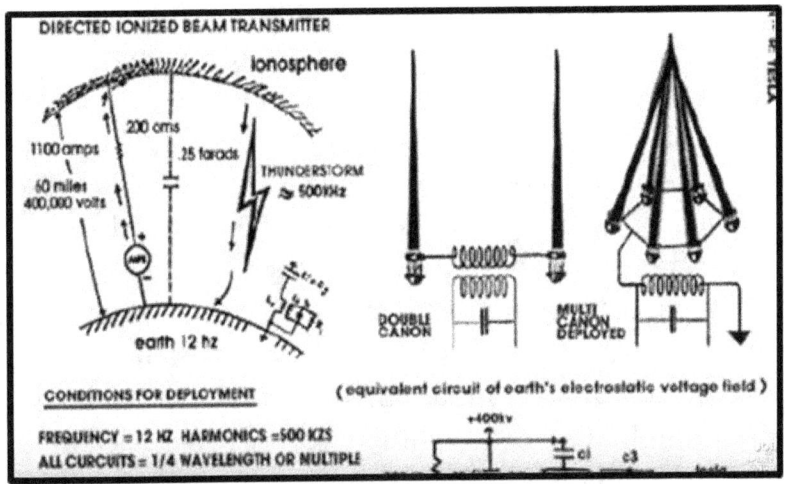

Second Event-A repeat of this extremely dangerous experiment would take place many years later. Throughout what was called the very secret *"Rainbow Project"*, Nikola would be skeptical. Later we will describe some of the effects of an application of this second event where Hugh, high frequency, Magnetic fields may have showed us time travel. It would be at a terrible price. We will look at that later. Right now let's hear about something sort of funny.

1897 Earthquake Machine

I know you are wondering how a machine that could cause earth quakes would be funny, but just wait. Nikola's machine cracked windows, shook girders of a building, and caused the police to be called. For a long time, this machine and its effect were considered part of fable but in the 1980s resonant frequencies on a bridge in Washington were self-amplified and soon the entire bridge came down. In the 1990s high frequency electromagnetic vibrations turned aluminum to a jelly-like substance easily pulled apart which could have made a building fall. In another experiment, a knife was driven through the metal with ease. We will describe this series of experiments later, but right now, Nikola and Mark Twain are the main characters. The time was somewhere between 1890 and 1897, in New York City. Nikola Tesla & Samuel Clemens, aka Mark Twain, were very good friends but that does not mean tricks were not played. The two of them were known to get into their share of trouble after a night at the gentlemen's club and Sam lived just a few blocks from Tesla's Laboratory at 35 South Fifth Avenue. Mr. Clemens visited Tesla's lab from time to time. Before we look a Mark Twain's encounter, let's read a first-hand report.

Reporter A.L. Besnson indicated he witnessed a test of a small vibrating device where a thick steel rod broke in half then he said, *Tesla put his little vibrator in his coat-pocket and went out to hunt a half-erected steel building. Down in the Wall Street district, he found one ten stories of steel framework without a brick or a stone laid around it. He clamped the vibrator to one of the beams, and fussed with the adjustment until he got it. Tesla said finally the structure began to creak and weave and the steel-workers came to the ground panic-stricken, believing that there had been an earthquake. Police were called out. Tesla put the vibrator in his pocket and went away. Ten minutes more and he could have laid the building in the street. And, with the same vibrator he could have dropped the Brooklyn Bridge into the East River in less than an hour.* Images below are said to be some variations of the original "Electrical Oscillator" that may have been a component of the Earthquake machine.

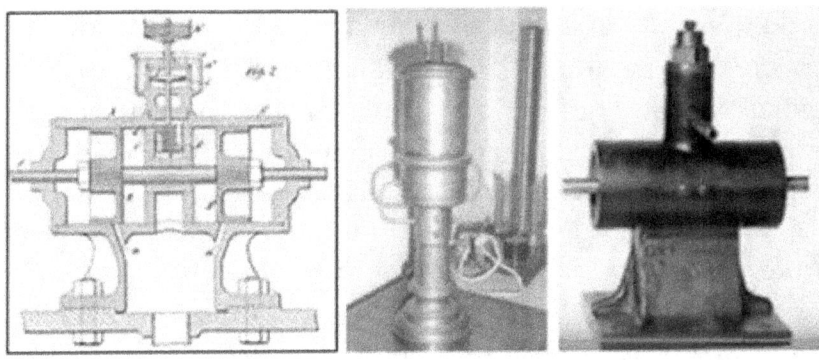

It supposedly shook his building and nearly brought down its walls. It was a high frequency oscillator that was powerful, but not enough to destroy his lab. So, he continued to test and many felt a shaking in New York enough to get over to Tesla's lab where he had destroyed is original model. In a later model a piston was attached underneath a platform. Within a short time, the table could be made to shake violently while witnesses were standing on it. Mark Twain was one such witness and he found out the shaking was therapeutic, but after a few minutes, <u>it acted like a strong laxative</u>. I know some of you are thinking about the incident in the 1960s where a research team used a certain vibration, believed to have been around 160 Hz in an ill-fated test similar to Tesla's.

Sphincter Resonance Test

While some have suggested this part of the Earthquake machine test was also made up, we travel to the 1960s. Scientists were using vibrational emissions to excite the brain rather than make earthquakes. Not everyone gained success as can be illustrated with something called Sphincter Resonance. In the 1960s, somebody discovered the resonating frequency of the sphincter [believed to be around 160 Hz]. A team created a device <u>later </u>called an "Anal Sphincter Resonator". It was, supposedly, kind of like a musical organ. The idea was to intensify the "suspense" in movies whenever "Danger" was about to be portrayed.

They had determined that around 4Hz vibrations would initiate confusion; 6Hz would kick in enhanced visual imagery; 13 Hz would establish a level of anxiety, and 28 Hz would establish a level of terror. [See the following table for details obtained from hundreds of tests.]

BACKFIRE and more BACKFIRE. Apparently, someone was experimenting with other frequencies, or the building beat frequencies caused an un wanted effect. The vibrations caused the entire audience to soil themselves. The specific group of tones generated by this contraption has been referred to as a 'Brown Note' for some reason that I am not going into at this time. The specific notes have been lost over time, so I'm sure one of these mishaps will occur again in the future or already happened to Samuel Clemens.

The following is a reduced list of some of the details we are now "RE-finding about vibrations of the mind. One such method is to use a device called a Transcranial Magnetic Stimulator but there are many devices that are modifications of Tesla's first joke of Mark Twain.

Type	Freq. (Hz)	Normal Reactions
Epsilon	<0.5	Extraordinary states of consciousness, High states of meditation, Ecstatic states of consciousness, High-level inspiration states, Spiritual insight, Out-of-body experiences, Suspended animation.
Delta	0.2 to 4 Hz	Confusion, boosting intuition, Deep sleep, Lucid dreaming, Increased immune functions, Hypnosis, Anti-aging, Increased intuition, Inner being & personal growth, Trauma recovery, Near death experience, Blissful "being" state
Theta	4 – 7 Hz	Arousal, Deep relaxation, Increased memory, Creativity, Hypnagogic state, Access to subconscious images, Reduced blood pressure, Profound inner peace, emotional healing, Inner wisdom, Faith, psychic abilities, Twilight sleep learning, Vivid mental imagery, Military remote viewing
Alpha	8 – 12 Hz	Relaxation, Meditation, Light relaxation, Positive thinking, Creative problem solving, Mood elevation, Stress reduction, Intuitive insights, Daydreams, Calm, relaxed, Lucid mental states, Tranquility, Detachment
Beta	12 – 30 Hz	Alertness, Anxious thinking, Active concentration Analytical problem solving, Judgment, Decision making, Increased mental ability, Focus, Good for absorbing information passively, Treating Hyperactivity, Sensorimotor Rhythm, Outer awareness, Arousal, Dendrite growth,
Gamma	30 – 100 +	Motor functions heightened, Boosted memory, Enhanced perception of reality, Binding of all senses, Increased compassion, High-level information processing, Natural antidepressant, Positive thoughts, Higher energy levels, Decision making in a fear situation, Muscle tension, Release of growth hormone, muscles, Recovery from injuries, Rejuvenation effects

61

Oil Exploration Device-Tesla claimed the device, properly modified, could be used to <u>map underground deposits of oil</u>. A vibration sent through the earth returns an "echo signature" using the same principle as sonar. This idea was actually adapted for use by the petroleum industry, and is used today in a modified form with devices <u>used to locate objects at archaeological digs</u>.

Fortunately for those trying to discredit Nikola Tesla, this and many more experiments were deemed impossible and dismissed even when eye witness accounts supported his claims. Unfortunately, this is the normal actions of scientists wanting to secure their inappropriate theory at the risk of holding science back.

Let's get back to Mark Twain. While his soiling himself might have been an aggravation, he and Tesla stayed close friends. The following image show Mark Twain participating in an experiment when thousands of volts could be sent through a person without harming them. Nikola is in the background standing by just in case action was needed.

From earthquakes Nikola went on to build a robot.

1898 Robotic Boat

Tesla publicly demonstrated his "automaton" technology by wirelessly controlling a model boat at the Electrical Exposition held at Madison Square Garden in New York City during the height of the Spanish-American War. To add a flare for showmanship, the boat was asked, *"What is cube root of 64?"* It responded by flashing a lamp 4 times. It was truly a robot boat.

Tesla used his radio remote control to change its course, and react to commands and everyone loved it. Tesla knew this would make the world safer in War, but the Government did not seem interested. The following images show the side shot of his unique device. While it did not have a balast pump to push water in and our of a chamber like a submaring, the entire ship was sealed and needed no openings for crew at all, so it could have been turned into a robotic sub. As part of the device and brand new type of electrical turbine was also designed. Its mechanical characteristics are shown to the right.

Other portions of the patent are shown below.

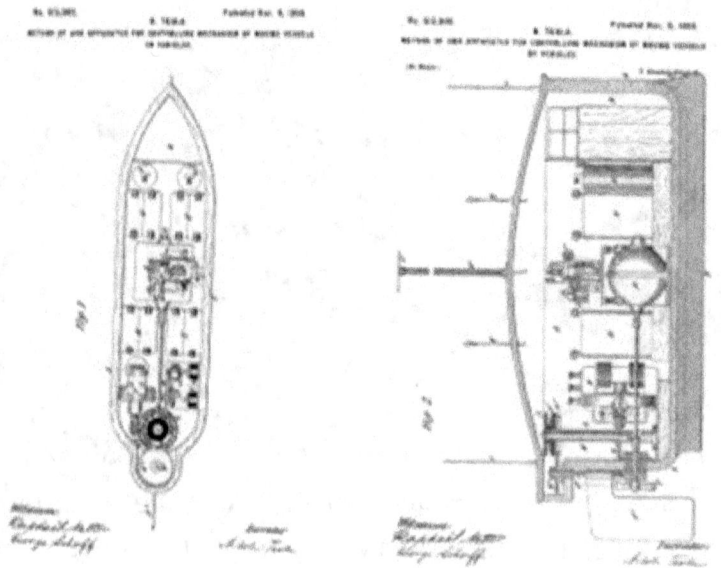

Near the bow and stern were two small metal tubes about a foot high surmounted by small electric lamps. The interior of the hull was packed with a radio receiving set and a variety of motor-driven mechanisms which put into effect the operating orders sent to the boat by remote control. There was a motor for propelling the boat and another motor for operating the servo-mechanism, or <u>mechanical brain, that interpreted the orders coming from the wireless receiving set and translated them into mechanical motions</u> including the following:

- steering the boat in any direction,
- making it stop, start,
- go forward or backward,
- light either lamp [Including answering complex math problems aided by the remote controller's knowledge.]

1898 Robotic Cars

But we are not going to stop at this. Tel-automata [robotics] will be ultimately produced, capable of acting as if possest of their own intelligence, and their advent will create a revolution. As early as 1898 I proposed to representatives of a large manufacturing concern the construction and public exhibition of an automobile carriage which, left to itself, would perform a great variety of operations involving something akin to judgment. But my proposal was deemed chimerical at that time and nothing came from it.

While Nikola had established the beginnings of all remote-controlled devices and robotics, he wanted to do more so he went to Colorado.

1899 Colorado Ground Current

Nikola built a massive tower and high voltage electricity laboratory near Pike's Peak and he finally made lightning. Sparks were everywhere, High voltage and high frequencies made the place sing and dance lightning. When the first experiments in Colorado Springs Experimental Station were performed, Tesla recorded his initial spark length at five inches long. He described it as *very thick and noisy.* During one experiment, the Colorado Springs power plant was destroyed and Tesla was forced to rebuild broken parts or they would not allow him to connect again.

Current Around The Globe

In his "*My Inventions*" autobiography Tesla stated, "*When in 1900 I obtained powerful discharges of 100 feet (in the Colorado Springs laboratory) and flashed a current around the globe, I was reminded of the first tiny spark I observed in my Grand St. laboratory and was thrilled by sensations akin to those I felt when I discovered the rotating magnetic field.*"

This discovery was highly criticized but Tesla received a signal 80-thousandths of a second after his initial transmission, so something happened. If this was so, his losses were very low to go around the globe. The following images show his 145-foot antenna and massive electrical flashes inside the experimental station.

Lightning Radar

As I was improving my machines for the production of intense electrical actions. I was also perfecting a means for observing feeble efforts. One of the most interesting results and also one of great practical importance, was the development of certain contrivances for indicating at a distance of many hundred miles an approaching storm, its direction, speed, and distance traveled.

1899 Outer-space Signals

Working late one night on his powerful and sensitive radio receiver, Tesla observed strange rhythmic pulses on the receiver. He concluded that there was no possible explanation other than some effort was being made to communicate with Earth by creatures from another planet. Tesla reveals the discovery and is highly criticized.

I felt as though I were present at the birth of a new knowledge or the revelation of a great truth. My first observations positively terrified me, as there was present in them something mysterious, not to say supernatural, and I was alone in my laboratory at night, but at that time the idea of these disturbances being intelligently controlled signals did not yet present itself to me. The changes I noted were taking place periodically and with such clear suggestion of number and order that they were not traceable to any cause known to me. I was familiar, of course, with such electrical disturbances as are produced by the Aurora Borealis, and earth currents, and I was sure as I could be of any fact that these variations were due to none of these causes.

The natures of my experiments precluded the possibility of the changes being produced by atmospheric disturbances, as has been rash;y asserted by some. It was some times afterward when the thought flashed upon my mind that the disturbances I had observed might be due to an intelligent control. Although I could not dicipher their meaning, it was impossible for me to think of them as having been entirely accidental. The feeling is constantly growing on me that I had been the first to hear the greeting of one planet to another. A purpose was behind these electrical signals. I have devoted much of my time during the year pasr to the perfecting of a new small and compact apperatus by which energy in considerable amounts can now be flashed through interstellar space to any distance without the slightest dispersion.

Could This be Real?

The message he received in 1899 by extraterrestrial communication was a simple 1, 2, 3. Someone had to have sent it. This would be the beginning of the Radio-telescopes in use today and the almost frantic desire for extraterrestrial contact. When Tesla first described it, like many of his other discoveries that were beyond normal comprehension, was met with ridicule and scorn. Some might think there cannot be real signals from near space because our moon, Venus and Mars are all dead planetoids, but is that a truth? The image below shows a tiny fraction of the details that appear to be from those who had lived or may still be living on the moon.

The first image shows what appears to be a small group of building and a massive sign. The second image shows a huge tower and a wall around the tower, interestingly, the top of the tow has a level floor for observers. Another similar tower is shown one the second row left and clearly visible lights in a square depression are shown. Finally, a roadway intersection at right angles and several square buildings are shown.

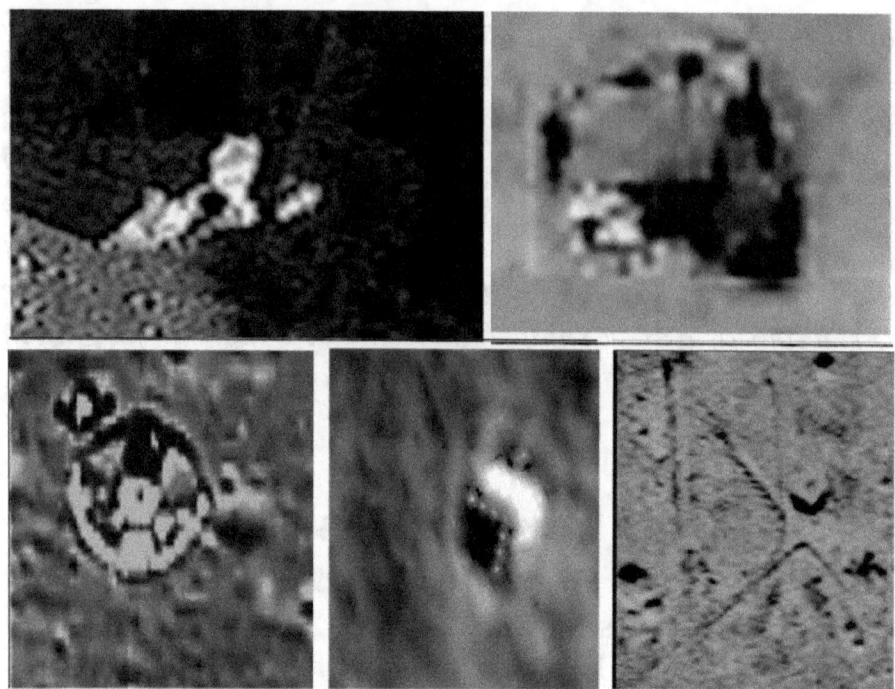

The following image shows and active vehicle on the lunar surface. Supposedly, no one is there. If they have vehicles and lights, they probably communicate with RF Waves that someone could pick up.

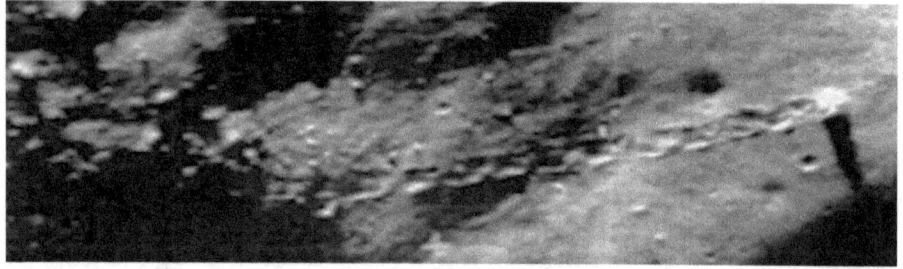

What about Venus?

Going to Venus is no longer reasonable and no one can live there today, but a number of buildings, roads, dried up riverbeds, and rolling hills tell us this planet once was a beautiful place and livable. [See the following collage of burned up buildings.

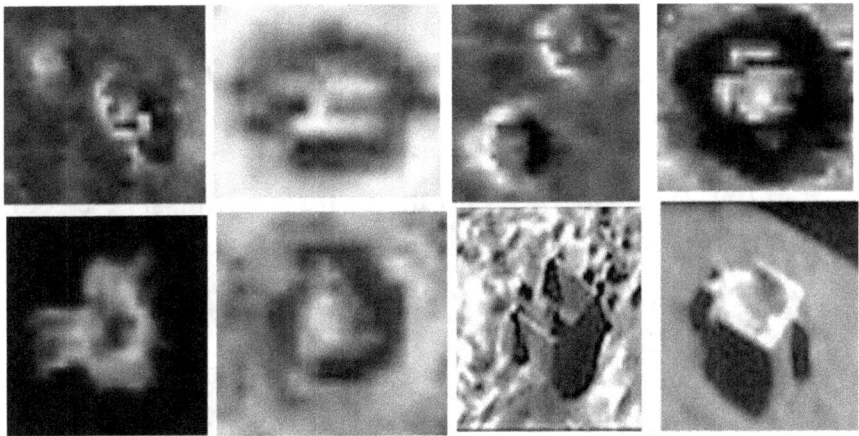

Tesla had not received communication from Venus as it is now 700 degrees and has a caustic, high pressure atmosphere and anyone who had lived there are certainly long dead. Not long ago, it appears there was a war taking place. In the following image we can see between a small town and a roadway stretching many miles are 7 perfectly sized and circular blasts that were caused by 7 identically sized things exploding. Some believe these were bombs dropped from the air.

What About Mars?

Around the planet we find indications of life including running water, vehicles travels, industrializatyion forestation, and a number of areas where underground tunnels reach the surfacfe. The first 2 images show one of the many industrial areas that have been located. Pleas notice th ebuilding with the 8 pillars marking its entrance and the massive circular towere to the left.

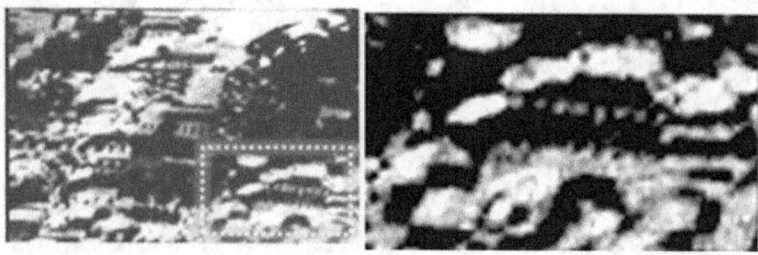

In another industruial area we can make out what appears to be a two story builing with windows and even window like areas on the roof.

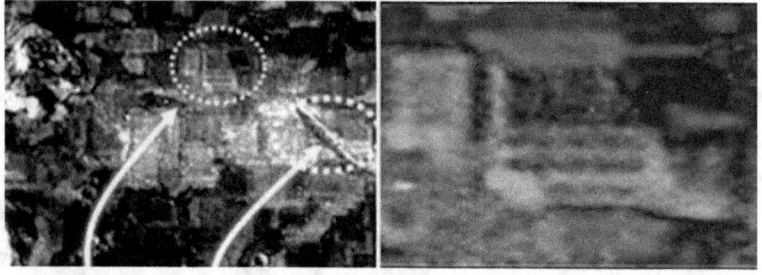

Besides having buildings that were once occupied, we find plants all over the place. This looks like someone is trying to terraform the planet back to the way it was 10 thousand years ago with breathable oxygen and an atmosphere that would allow for surface water to be in abundance. Below are just a few of the sights with vegetation of some type. Presumably these are making oxygen.

There are also many visible tunnels that are partially surfaced today that shows underground life was, or is still viable. Below are just a few. One had been damaged as shown.

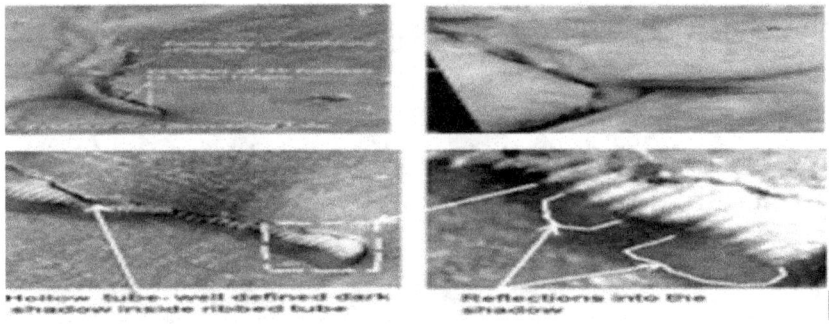

On the surface we are now finding evidence of discarded vehicles, flying vehicle and ground vehicles. Many of which are still active as shown.

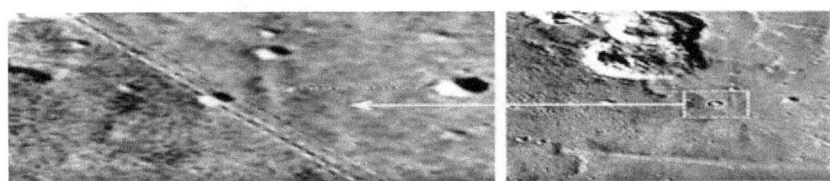

1900 Tesla's Ozone Company

People were laughing so Tesla developed a new enterprise. In 1896, Tesla was issued a patent for a corona discharge ozone generator using charged metal plates to act on ambient air. He formed the Tesla Ozone Company and went into production of these units. His customers were naturopaths and allopaths who welcomed this powerful therapy into their practices. Breathing of ozone bubbled through olive oil and other oils was widely practiced at this time, and the Sears catalog of 1904 offered a unit for this purpose using eucalyptus, pine and spearmint oils. Tesla produced a gel made by bubbling ozone through olive oil until it solidified, and sold it to doctors.

Tesla Coil
(circa 1891)

Water Treatment systems- Compact coil designed for use as an ozone generator for water treatment.

This time was troubleing for Nikola as he began looking like what Hitler would eventually look like. Some believe Hitler made himself look like Tesla to make himesel look smart. We cannot blame Nikola as his look was before Adolf hitler was in power. In the images below, the last one is Nikola and the first 2 were images of Hitler. Luckily he had a mishap with his shaver and his mustash became shortened.

1901: Turbine and Ocean Energy Generator

By 1901 Tesla had developed a new type of turbine that had no fins to slow operation, reduce efficiency, and increase sound levels. Some examples are shown below, but he looked for ways to use this turbine. He would use it in later flying machines, but right now he was looking to lock into the energy of the Ocean.

Ocean Power-Another of Tesla's ideas was to enable the production of cheaper power based on the utilization of the temperature differences. While that was the basis of heat pump operations, Nikola concentrated on the oceans. He noted that between the ocean surface water and that of water three miles below the surface or in some cases significantly less deep there was a MASSIVE thermal gradient. The basic idea of heat pump operations was not a brand-new idea, but

Tesla devised an interesting engineering scheme which makes the idea practicable. The following sketch shows how the icy waters from the ocean depths are brought in contact with the considerably warmer waters of the upper levels. The differential could cause a steam that could great turbines to make electricity.

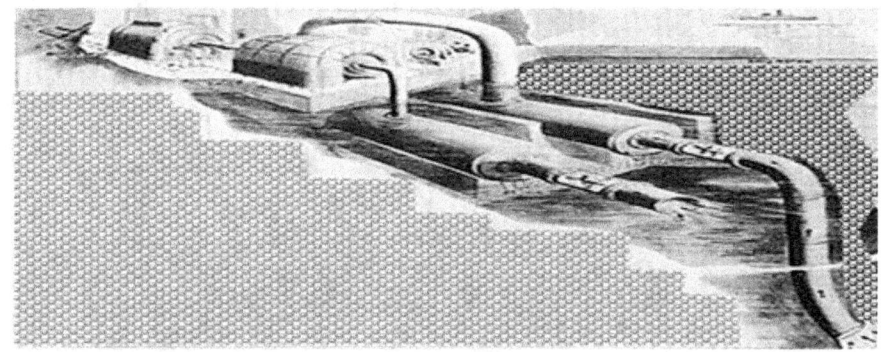

From his pattern we find the following: *"Plan of a system whereby the transfer of vapor between two vessels at different temperatures drives the armature of electrical generator."* [Below left] *"Here the water, or other fluid operating the turbine D is kept in a closed system, circulating through condensers immersed in water of different temperatures."* [Below middle]

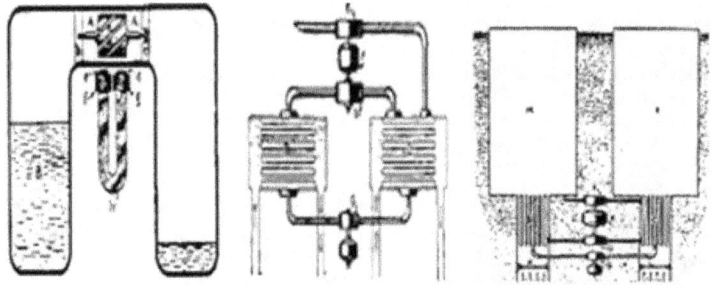

The basins H and I are filled and emptied by the tide, saving much of the energy otherwise expended in pumping. [preceding right]. All this is great and many of the tidal green energy devices still use Tesla's example for production of energy, but Nikola went a step farther.

The following is the design of a vessel to be propelled by energy derived from temperature differences in the water. The symbols designating the operating mechanism are the same as that explained previously, but now we have a completely mobile self -energy producing sea vessel.

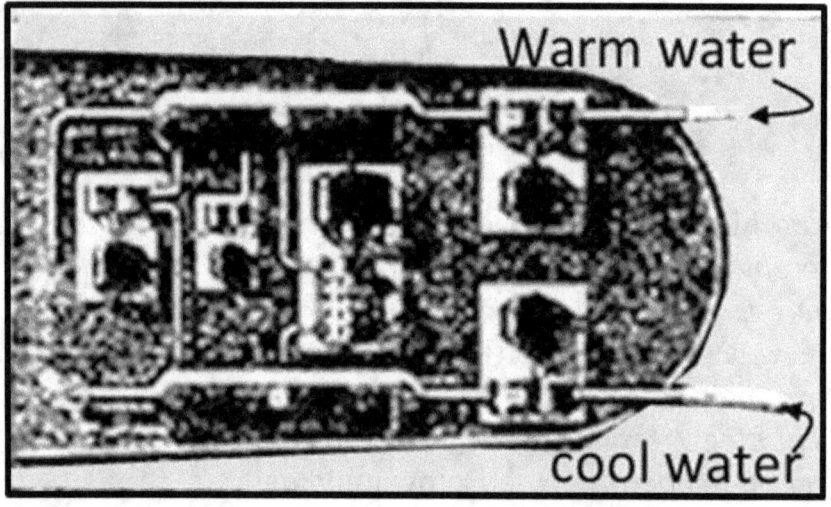

From free energy boats and energy form the oceans, Tesla changed directions and looked at solar energy.

1901 Solar Energy and Neutrinos

In 1901 Nikola Tesla was one the first to identify "radiant energy." Tesla says that the source of this energy is our Sun. He concluded that the Sun emits small particles, each carrying so small of a charge, that they move with great velocity, exceeding that of light. Tesla further stated that these particles are the <u>neutron particles [now called Neutrinos]</u>. Tesla believed that these neutron particles were responsible for all radioactive reactions. Radiant matter is in tune with these neutron particles. Radiant matter is simply a re-transmitter of energy from one state to another.

Tesla Describes the first solar Charging System

Stick an antenna up in the air, the higher the better, and wire it to one side of a capacitor, the other going to a good earth ground, and the potential difference will then charge the capacitor. Connect across the capacitor some sort of switching device so that it can be discharged at rhythmic intervals, and you have an oscillating electric output. His simple solar energy collection system is shown below left.

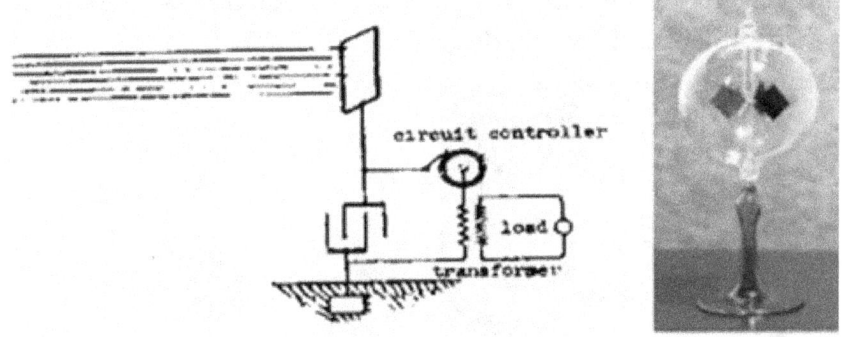

The patent refers to "*the sun, as well as other sources of radiant energy, like cosmic rays,*" that the device works at night is explained in terms of the night-time availability of cosmic rays. Tesla also refers to the ground as "*a vast reservoir of negative electricity.*" Tesla was fascinated by radiant energy and its free-energy possibilities. He called the Crooke's radiometer [previous right], a device which has vanes that spin in a vacuum when exposed to radiant energy *"a beautiful invention."* He believed that it would become possible to harness energy directly by *"connecting to the very wheel-work of nature."*

Tesla Found Neutrino emissions in the form of Cosmic Rays-*And as for the cosmic ray: I called attention to this radiation while investigating Roentgen rays and radioactivity. In 1899 I erected a broadcasting plant at Colorado Springs, the first and only wireless plant in existence at that time, and there confirmed my theory by actual observation. My findings are in disagreement with the theories more recently advanced. I have satisfied myself that the rays are not generated by the formation of new matter in space, a process which would be like water running up hill. According to my observations, they come from all the suns of the universe - Some of these rays are of such terrific power that they can traverse through thousands of miles of solid matter".*

We know now about 65 million neutrinos from sun go through just one of your fingernails every second. So, we are talking about a massive source of energy to be tapped and Tesla discovered them. This number goes up markedly during solar storms like that shown below left. Neutrinos are the disintegration of atomic nucleus as shown. As an electron is expelled, a neutrino must also be produced.

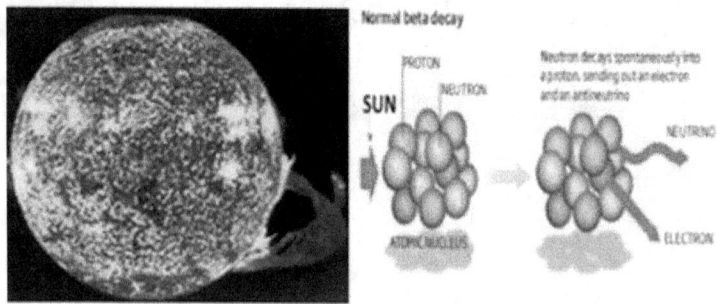

One might think this energy could be harnessed. While he was considering Cosmic Rays and Neutrinos, he began developing another massive tower. This time it would be in New York.

1903 Wardenclyffe Testing Begins

Tesla got seed money for J.P. Morgan to build a different type of energy plant drawing energy from the sky and allow transmission of radio signals around the world. Little did he know that Tesla was not only trying to pull in energy from the Ionosphere and possibly allow for RF transmissions, but also to push energy out through resonant modulated ground currents to make electricity free to everyone. Construction finally began for Tesla's most ambitious project, The World-Wide Wireless Power System known as Wardenclyffe Tower. The image below left shows the completes massive tower. Beside it is the remains after it was blown up in 1917 to insure the capability would not fall into the hands of our enemies. I know that sounds stupid, so let's investigate a little.

This was one of Telsa's more infamous projects. The Wardenclyffe Tower stood 187 feet tall, and was located on

Long Island, New York. Tesla envisioned it as both a means for facilitating world-wide wireless communication, a hundred years before cell phones would become popular, and as a method for delivering electrical energy over great distances. Tesla believed he could transmit both radio waves and electric power between continents, and by eliminating the need for wires. He proved this on a much smaller scale in his experiments, using his famed Tesla Coil. With the Wardenclyffe Tower he believed he could show the results of his work to the world. The plan was to build similar towers in major cities around the world, enabling energy and information to be transferred from point to point.

Unfortunately for Tesla, his financial backers left him before he could finish his work on the Wardenclyffe Tower. The general details of the tower form his patent are shown below. Essentially, it was a huge resonant tuned tesla coil. The coils produced millions of volts which ionized the air and allowed electrical discharge between the Ionosphere and the tower in the form of lightning, but it was much more. Nikola had discovered that if he adjusted the resonant components of his devices just right, the resistance to current would be significantly reduced. At very high frequencies, one could build a modulator with sufficient enough Q [ability to hold a single resonant frequency] and massive amounts of electricity could be sent both in the air and through the ground.

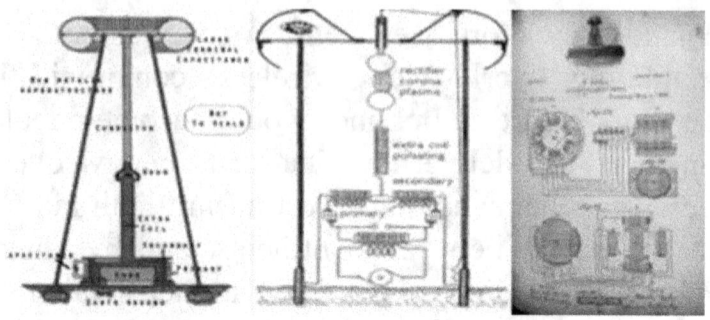

As he began his initial testing, Nikola said the worst thing he could say, as far as J.P. Morgan was concerned. "Electric power is everywhere present in unlimited quantities and can drive the world's machinery without the need of coal, oil, gas, or any other common fuel. This would hurt the profits of many powerful individuals including J.P. Morgan, John Rockefeller, Paul Warburg, and even a man named Thomas Edison. Morgan not only pulled funds from the project, but a massive campaign was initiated. They pushed newspapers, they owned, to smear Nikola as a charlatan so that no one would be unable to trust him enough for them to provide him money needed to continue. By 1908, his project was bankrupted. This bankruptcy may have been the catalyst for an unfortunate accident we will look at later. Just remember 1908.

"The New York Sun" reported strange effects at Tesla's transmitter about this time. It stated, *"All sorts of lightning were flashed from the tall tower and poles,"* and *"the air was filled with blinding light."*

8000 BC Ancient Power

Similar to the Warenclyffe tower in purpose, from a number of ancient texts and substantial physical evidence, we find that the machine called the Great Pyramid of Gaza was built by a man named Thoth. He built it according to various records over 10 thousand years ago. When he explained why he built it, he sounds like Tesla but one thing is for sure, the crystal described in Thoth's writing is supposed to draw out the force from <u>the Aether</u>. Before we get into how, let's read his own words.

Thoth's Writings- *Know ye that in the pyramid I built are the Keys that shall show ye the Way into light. I built a doorway leading down to the source of light. Then I raised over the passage, a mighty pyramid, <u>using the power that overcomes gravity</u>. Inside I carved a circular passage reaching near the top. There <u>I placed the crystal, that could draw the force from the Aether</u>. In the pyramid, <u>I built my knowledge of "science" so that it might be here when I am reincarnated.</u> - Seek thou in my pyramid, deep in the passage that ends in a wall.*

Apparently, Thoth made the pyramid to pull electricity out of the Earth rather than doing it with magnetics as Tesla was doing. Thoth made his machine use pyro-electric **crystals** in massive Red Quartz slabs. If you have seen the old Kodak "Magic cubes", you know what I'm talking about. They used a Crystal. When the crystal was crushed, it made a spark and as the pressure was released it made another electric spark. Each time a strobe would flash to take pictures without a battery. One could do this a million times and it would still make a small amount of electricity. All Thoth had to do was make everything bigger.   While this version is an extremely shortened description laid out completely by an engineer name Christopher Dunn believe me when I tell you it has been well investigated for reasonableness.

Electricity Pyramid

Here is where we find the "electro" for the ancient electroplating, electric machinery, electric heating, electric motors, electric lighting, and all sorts of things we find evidence of, around the world, with no previous explanation. Running fancy genetic laboratories, and the operation of complex flying machines was accomplished before the great Bharata War 5000 years ago. While we know that all these thing happened, the question we must face is—"how?". Thoth knew the secrets of Undall including the generation of massive amounts of electricity so he made the Egyptians a massive power plant [drawing the force from out of the Aether]. I know it will be hard to believe but stay with me for a little before rejecting this notion. We call the electric plant "the Great Pyramid" and we can be sure that it was working before the worldwide flood and was still operational for many thousand years after the flood. It, most likely, produced almost free power that could be supplied to remote sites without wires. The image following shows what is called the

87

Great Pyramid was actually the Great Electrical Power Plant. I know you have heard how the great Nickola Tesla almost completed his free energy electric plant in the early 20th century, before JP Morgan pulled his funding to halt the generator, but Thoth had no such money worries. The following sketch shows what are believed to be critical elements of the pyramid Electric Plant.

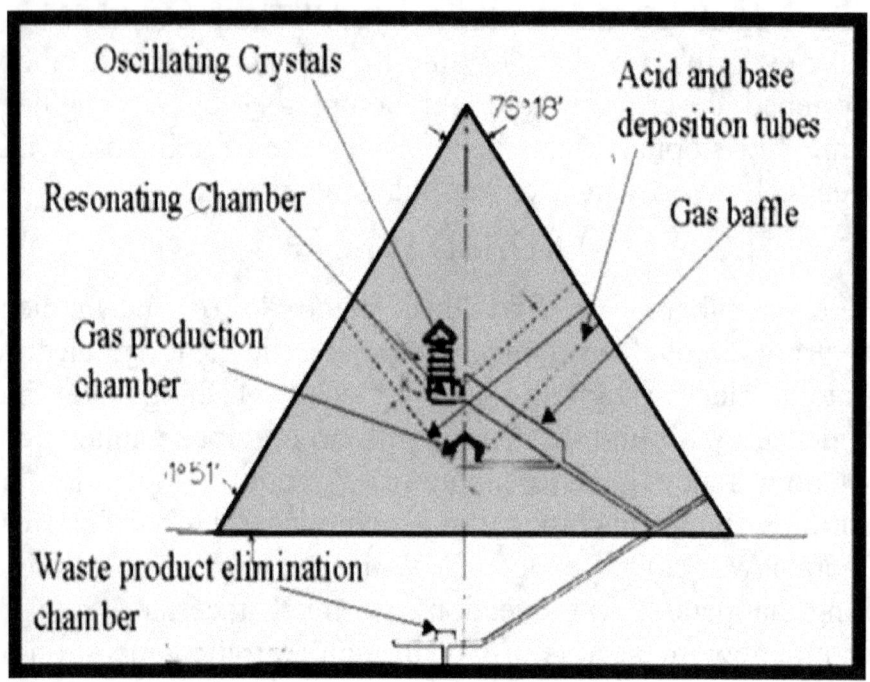

I know that sounds crazy, so I'm going to have to discuss the theory behind the crazy statement or this amazing element of history will only sound like an absurdity. If it didn't sound absurd at the top of the page, please understand; I'm talking about getting electricity without wires from a bunch of blocks. [Just thought that I'd emphasize the apparent absurdity before we examine the evidence and the modern experiments--crazy as it sounds; modern experiments strongly suggest this probability.]

Real Pyramid Power-Finally, someone came up with a plausible reason for Thoth making the Great Pyramid.

- It explains why it was made to such exact dimensions;

- Why all the different internal chambers were designed,

- Why the high crystalline content, granite boulders were necessary,

- Why massive granite slabs were polished where they could not be seen,

- Why the creators made the tiny shafts that were so difficult manufacture at angles THROUGH the other building blocks.

Christopher Dunn initially theorized and has tested his generator theory with positive results. His theory states that the great pyramid could very well have been used to produce large amounts of electric energy. Please believe me when I tell you that the occurrence of all of the anomalous elements associated with the Great Pyramid were not happenstance. The ancient people went to a lot of trouble to make the great pyramid the way it was.

The Great Pyramid is loaded with unbelievably accurate shafts cut into and through limestone at oblique angles and granite slabs are critically placed and polished where the polishing cannot be seen. Added to this strangeness was the discovery that the chamber dimensions show remarkable resonance levels, which we will address in a minute. Recently, we found that at least one of the miniature shafts had been sealed up under the orders of some ancient pharaoh. The great pyramid also has five, high crystal content, polished, granite slabs positioned on top of one another but spaced apart so that they can vibrate and below them is located a very interesting chamber that shows the signs of explosions. All of these unusual features were put into the pyramid when it was built, because it was built for a purpose different than holding a dead body.

Production of Gas-Today we know that mixing an "acid" and just about any "base" together forms hydrogen gas. We also know that pure hydrogen is extremely volatile and will cause an explosion if it meets air and a small amount of heat or spark. After the reaction is completed, the hydrogen would have mixed with oxygen in the air and would have produced water. The remains of the acid/base mixture that was left after the gas was produced would have turned into a salt

Salt Bi-product-Did I mention that salt deposits have been found in three places in the great pyramid? Large quantities have been found in the "queen's [gas production] chamber", and smaller quantities have been found in both the "grand galley [gas baffle]" and the "king's [resonating] chamber".

Electricity from a Crystal-Let me get back to some basics again. Today we constantly use the fact that when quartz crystals, such as those found in granite, are compressed, they produce electric sparks. When the pressure is relieved, the crystals produce more electricity. We also use the fact that a crystal will deform itself at a specific cyclic rate that is dependent on the cut of the crystal and its dimensions whenever excited by electricity. This whole crystal power thing and placing a certain crystal on your body to change your mood, relationships, or wealth may be crazy, but the power in all crystalline structures noted above has been used in electronic equipment for many years.

Electrical System-It looks like the very ancient Egyptian rulers from Atlantis knew these things and made a gas manufacturing, granite crystal compressing, electricity producing, machine which we call the Great Pyramid. A schematic of the pyramid as an electricity generator, which shows the component parts, was pictured earlier. The names given are not the same names we have been told, but these are most likely the functional names.

Resonating Chamber and Oscillating Crystals-As discovered recently by researchers Rick McCullum and Bill Cox, almost everything in the "King's Chamber" resonates at 640 Hertz. Five polished red granite crystals are placed above the "king's chamber and are separated from one another. My bet is that the crystalline mass making up each granite block above the "King's Chamber" has a resonance of 640 hertz. The granite blocks were not just found to be 640 hertz, but were "tuned" like we tune crystals today, by polishing them down to the precise dimension required.

Once the chambers, shafts, resonating crystals and properly sized resonance cavities were in place, all the Egyptians had to do, was find a way to continuously compress the crystals enough to produce electricity. Someone thought about it for a little and might have said, *"The pyramid will compress the granite crystals by itself."* Sure enough, the weight of the pyramid pushing on the granite caused electric energy to be output from the crystal. It wasn't some magic and they could not have kept the crystals from producing the initial surge of electricity. The dynamic elements of the machine are shown below.

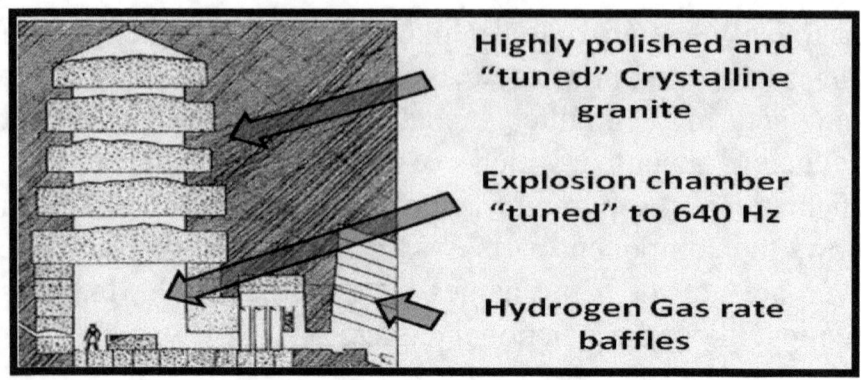

Highly polished and "tuned" Crystalline granite

Explosion chamber "tuned" to 640 Hz

Hydrogen Gas rate baffles

Instantaneous and Continuous Electricity-The crystal, like all crystals would resonate and begin to compress and expand itself at its base frequency. Typically, these "oscillations" die down very quickly, but simply having this great pressure on the blocks would produce several "cycles" of electricity. Of course, the electricity had nowhere to go so a very high "voltage level" would be reached and like miniature lightning bolts, electrical sparks would be generated. Then the energy would dissipate as heat and go away.

Resonators-This "short lived group of electrical sparks" would need to be further amplified so something called "resonance" was added in the Pyramid. Resonance is a property that amplifies or sustains one frequency of electromagnetic energy while ignoring other frequencies. This property of resonance was initiated by first having the granite slabs polished to the same basic dimensions and was further extended by making the "room" to specific dimensions. Finally, the resonance was enhanced by building sort of a tuning fork or vibration cavity at the center of the room. Some may tell you that this vibration cavity is a sarcophagus, but it doesn't make sense. There is no lid and no evidence that a body had ever been placed in the tuning fork. By the way, <u>the frequency that all of these elements seem to amplify is 640 Hertz</u>. Even the lip of the "vibration cavity" was possibly rounded smooth to insure optimum sustainment of oscillation. This is no sarcophagus. Why in the world would the sides have rounded tops so that a non-existent lid could never stay secure?

Gas Production Chamber-All that was fine if you only wanted one burst of electricity, but sustainment was also planned. Hydrogen gas was, most likely, produced in the queen's chamber by combinations of materials transferred down the small deposition shafts visible today. Some "acid" would be poured down one shaft and some "hydrated base" went down the other. The volatile gases produced by the mixture would slowly rise through the baffles [sometimes called the Grand Gallery] and finally reach the "King's chamber" before the "Granite Crystal Oscillations" stopped. Sparks associated with the electricity produced by the granite crystals would ignite the gases and cause the chamber to, momentarily, get larger from the explosion. This action relieved the pressure on the Granite slabs, which, in turn, produced more cycles of electricity. Very quickly, the pressure of the pyramidic weight would take over and begin to crush the granite slabs once again to produce more cycles of electricity over again. Like an almost imperceptibly moving engine, the electricity would continue to be produced as long as hydrogen gas was allowed to enter the "Resonating Chamber". I don't mean small amounts of electricity either. It produced large amounts of electricity.

Explosion Evidence-Researchers have discovered that many of the boulders that make up the king's chamber have been moved out slightly and one of the granite slabs has even cracked due to the effects of these cyclic explosions so there is good evidence to support the oscillating cavity theory. Did I mention that the bottom of the lowest granite ceiling slabs is covered with a fine black dust as if some type of high rate burning process had blackened it? This blackening is found nowhere else in the pyramid.

Starting the Pyramid-After the "queen's chamber" had been filled with the acid and base mixture, and the hydrogen gas had tunneled through the grand gallery to the king's chamber; a flame was introduced through one of the openings to the chamber and the ensuing explosion began the production of electricity. The image following shows possible high voltage emission from the pyramid.

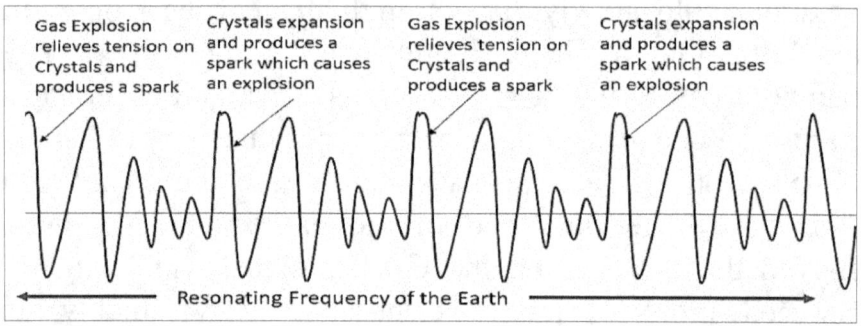

Sustainment-All the Egyptians had to do was to periodically refill the materials for the gas production and the machine continued to output electricity at about 640 Hertz. You might wonder, "What was so special about 640 hertz?" Nikolas Tesla may have discovered one part of the answer.

During the time Thoth first built this electrical generator for everyone to enjoy electrical power, just like Tesla; the earth resonance was much faster than it is today [about 640 Hz.] This frequency is extremely important in that the ground quickly dissipates electrical signals in the form of heat as it shows a huge resistance to operation. *If you transmit at the earth resonance, you can transmit the energy anywhere without loss*. We will talk about nuances of this power distribution in a minute, but first let's get back to Nikola.

1908 Earth Resonance

Not only did Nikola Tesla discover alternating current, but he also discovered that the ground **could** carry electricity. No electric wires were needed to carry electricity as most people believe today. He demonstrated to many this new capability and, reportedly, lit large quantities of electric lamps, which were placed at long distances away from his generator. He, somehow, used the earth as the conductor and lit them without wires. The problem with his newfound distribution method was that no one could control who was using the energy, therefore, no Earth transfer systems were ever commercially produced. No one would gain a profit and Tesla's method died with him.

Ground Current for Electricity-I know that you are probably saying to yourself that conducting electricity through the earth is impossible, but now we know it is not impossible, and we know it is not that simple but Tesla did it.

 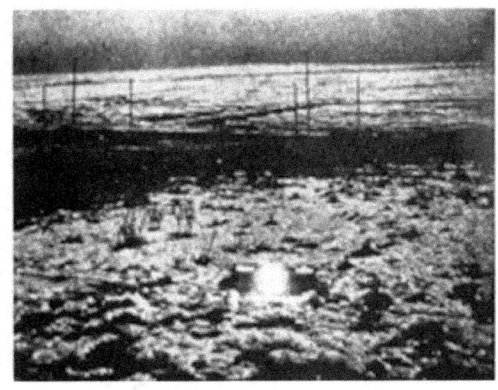

In 1899 at Colorado Springs Nikola transmitted high frequency high voltage electricity and then set up a ground tuned coil and light bulbs into the earth around the tower. The light bulbs lit up, up to one 100 feet away. In a second experiment he produced something like 100 million volts of high-frequency electric power wirelessly over a distance of 26 miles at which he lit up a bank of 200 light bulbs and ran one electric motor with a souped up version of his Tesla coil, Tesla claimed that only 5% of the transmitted energy was lost in the process but I personally think it was more. The method he would use to produce this wireless power was to employ the earth's own resonance with its specific vibrational frequency to conduct AC electricity via a large electric oscillator.

The entire energy of the current is transmitted through the ground exactly as though it went through a huge wire, Tesla said and added, *"In his experiments in Colorado it was shown that a very high voltage signal developed by the transmitter traversed the entire globe and returned to its origin in an interval of 84 one-thousandths of a second, this journey of 24,000 miles happened almost without any loss of energy."*—*"The practical applications of this revolutionary principle have only begun. So far they have been confined to the use of oscillations which are quickly damped out in their passage through the medium."* Tesla had found the resonance of the Earth itself.

Confirmed Earth Conduction

It really wasn't until 1976 that Dr. Andrija Puharich was able to point out that Tesla's power transmission system could not be explained by the laws of classical electrodynamics, but, rather, in terms of relativistic transformations in high energy fields. It seems that when an electron encounters a positron, the two particles would annihilate each other. Because energy can't be destroyed, the particles are transformed into an electromagnetic wave. Transversely, if a huge electromagnetic wave is manufactured, electrons and positrons of equal quantity must be manufactured. This recombination can be done at a remote site. Let me put this whole thing in the frequency domain and associate it with Vibrational matter.

All one would have to do is to somehow make the electricity go out of phase with this universe and it would no longer exist to the Earth. At a destination, simply relink the electricity to be visible to the Earth by converting the phase back to what it should be.

Ordinary electrical currents do not travel far through the earth. Dirt has a high resistance to electricity and quickly turns currents into heat energy that would be wasted. With this "pair- production" method, electricity can be moved from one point to another without really having to push the physical particle through the earth. The transmitting source would create a strong field, and a particle would only be created at the receiver by means of some catalyst. Without the catalyst, there would be no loss of energy and no heat would be generated. Put another way, the earth could sort of store the electricity until needed.

Back thousands of years ago- Evidently, the ground current distribution of electricity during the ancient times was used for many years and anyone that held the secret to the recombination process could tie into the system and pull out electricity without wires. Limiting the number of people who held the "secret" of reinstituting it must have controlled payment for the electricity. I know this sounds all mystic, but luckily, a researcher and structural scientist named Christopher Dunn will make it all make sense.

Pyramid Resonating Dimension-Like the resonance of the "king's chamber", researchers have determined that the base length of the Great Pyramid equals the distance that a sound wave at 640 hertz would travel in one cycle. This critical dimension, probably allowed the transfer of the 640-hertz AC electricity from the pyramid. Its resonance may have been very important in the transfer of energy through the earth just like Tesla had demonstrated in the form of a frequency that was a critical harmonic of the earth's normal vibration frequency. Once electricity was received, it would have to be stored inside a building with extremely high dielectric or the energy would all escape and no one could use the electricity for lights and machinery. We can believe receivers would use something like sheets of mica because of its massive insulating properties.

Mica for Electrical Insulation-Speaking of PreMaya, living in Mexico thousands of years ago, we find something very strange near Mexico City. We find a Pyramid with almost the exact same base dimensions as the "Great Pyramid" and internal to its structure we find conduits that seem to allow wirings, but the most unusual element is there are massive sheets of Mica in the floor and walls of the structure. As I said, mica is a great electrical insulator, but the only place the PreMaya could have gotten the stuff was in Brazil, so it was very important as supply would have been very expensive in time and money. Instead of being an Electric generator, it is as if this mica insulated Pyramid was used to hold massive amounts of electrical energy and secure it from discharge until needed. The mica lined Pyramid is called "Pyramid to the Sun" and it is substantially different than all other pyramids found in the Americas or anywhere in the world. Like the Great Pyramid, there are no writings on the walls and it is more massive than most. The pyramid is in the city now known as Teotihuacan but it was built well before the Maya, Aztec, Olmec, Zapotec or other known civilizations built the other structures around the strange building. The images following show the pyramid and a comparison to the Great Pyramid in Egypt.

Underneath the pyramid, a 12-foot-wide corridor was found and photographed by the robotic camera. The image taken shows a narrow, open space left after the tunnel was intentionally closed off about 2000 years ago and then filled with debris. The remaining portion of the passage way is more than 100 meters long and was excavated in the rock perfectly.

In some places tool marks can been seen showing that people once visited or <u>lived underground here</u>. Also, Mica was found in the walls. The tunnel is shown next right while the other 2 images are of different sections of the Mica flooring in th e pyramid.

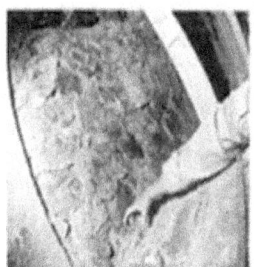  

In 1909 Tesla Revealed the Frequency of the Earth-As he tries to continue his efforts, Tesla explains what he already knowns and what he needs to do. *"This is not merely a theory, but a truth established in numerous and carefully conducted experiments. When the earth is struck mechanically, as is the case in some powerful terrestrial upheaval, it vibrates like a bell, its period being measured in hours. When it is struck electrically, the charge oscillates, approximately, twelve times a second. By impressing upon it current waves of certain lengths, definitely related to its diameter, the globe is thrown into resonant vibration like a wire, stationary waves forming, the nodal and ventral regions of which can be located with mathematical precision. Owing to this fact and the spheroidal shape of the earth, numerous geodetical and other data, very accurate and of the greatest scientific and practical value, can be readily secured. Through the observation of these astonishing phenomena we shall soon be able to determine the exact diameter of the planet, its configuration and volume, the extent of its elevations and depressions, and to measure, with great precision and with nothing more than an electrical device, all terrestrial distances. By proper use of such disturbances a wave may be made to travel over the earth's surface with any velocity desired, and an electrical effect produced at any spot which can be selected at will and the geographical position of which can be closely ascertained from simple rules of trigonometry".*

***Ground Become Wire**-"This mode of conveying electrical energy to a distance is not 'wireless' in the popular sense, but a transmission through a conductor, and one which is* *incomparably more perfect than any artificial one. <u>All impediments of conduction arise from confinement of the electric and magnetic fluxes to narrow channels</u>. The globe is free of such cramping and hinderment. It is an ideal conductor because of its immensity, isolation in space, and geometrical form. Its singleness is only an apparent limitation, for by impressing upon it numerous non-interfering vibrations, the flow of energy may be directed through any number of paths which, though bodily connected, are yet perfectly distinct and separate like ever so many cables. Any apparatus, then, which can be operated through one or more wires, at distances obviously limited, can likewise be worked <u>without artificial conductors,</u> and with the same facility and precision, at <u>distances without limit</u> other than that imposed by the physical dimensions of the globe. Millions of such instruments can be operated from but one plant of this kind. More important than all of this, however, will be the transmission of power, without wires, which will be shown on a scale large enough to carry conviction"*

In his patent, Nikola added the following. "*Mr. Nikola 'Tesla has announced that as the result of experiments conducted at Shoreham, Long Island, he has perfected a new system of wireless telegraphy and telephony in which the principles of transmission are <u>the direct opposite of Hertzian wave transmission</u>. In the latter, he says, the transmission is <u>affected by rays akin to light</u>, which <u>pass through the air and cannot be transmitted through the ground</u>, while in the former the <u>Hertz waves are practically suppressed and the entire energy of the current is transmitted through the ground exactly as though a big wire</u>.*

103

1908 Death Ray

The project was to be scrapped. The setback, and stigma of failure that surrounded that possibility, would have dealt a serious blow to Tesla and his self-esteem. So, he began investigations in a different direction. Tesla began pushing his brain to figure out how to war and began describing what he called "Teleforce". Some of his words were pointed and the newspapers called it "The Death Ray". Shown in the graphic, it was a modification of his tower where particles would be pushed with magnetic pulses until their speed was tremendous and be shot out in a direction by some type of magnetic focusing.

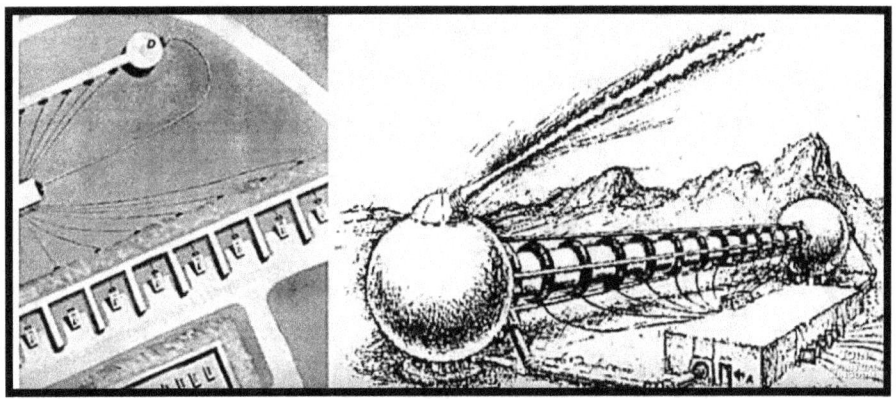

Teleforce [Death Ray]

To be clear: Teleforce, according to Tesla, was envisioned as a means of defense, never as an offensive weapon. He intended this "Death Ray" as protection, and as a method for eliminating war altogether. It was just as often referred to as the "Peace Ray" but the "Death Ray" caught on. It was in between a rail gun and a particle accelerator.

Tesla's mechanism of delivering a charged particle beam held the power, he said, *to wipe out entire armies and bring down 10,000 enemy aircraft at a distance of 200 miles.*

Teleforce was an incredible device with mind-boggling implications. These were the days before World War II, and we can believe that if Tesla's invention been put into practice thousands and thousands of lives might have likely been spared, but if the wrong people got the details, which side would have used it first? One of the articles is shown next.

1908 Tunguska Tragedy

The question some have asked is, "Did Tesla test his particle gun with his tower?" The details of his tower are not completely known and acceleration fields could very well have been used to generate the energy. If so he could have had the equipment to test his theory. With the Map showing how Tunguska lies on a direct line from Tesla's Wardenclyffe tower past Peary's 1908 location near the North Pole. Some have suggested a massive test to be witnessed by the respected Explorer went terribly wrong and the impact point could have been overshot.

Tesla may have invented a system for transferring massive amounts of energy through the air over hundreds or thousands of miles. So far this is interesting in theory, but what's the connection between Tesla and Tunguska?

As the story goes, explorer Robert Peary was undertaking an expedition to the North Pole around the time of the Tunguska Event.

Strangely, Tesla contacted him before the trip and asked him *to report back on anything unusual he encountered.*

Did Tesla then fire a blast of energy at the uninhabited North Pole and miss, hitting Tunguska instead? It seems extremely reckless that Tesla would have attempted a demonstration this way, but Tesla's experiment may have been the act of a frustrated and desperate man. Remember, the financial backing for his work was drying up and he already had this massive tower and massive magnetic field generator in his back yard and the last time he had gone to the military to show how to fight World War I with robots, they ignored him. He could believe the only way to protect Americans would be to convince the military to go forward with his device. Tesla never told anyone about this possibility, but, how could he?

Let's think about this a minute:

Some claim Wardenclyffe was not designed to be a mere communication system and already had the acceleration devices in it to use it for directed rays. Possibly this was even a requirement to direct beams of energy from point to point across the globe.

Tesla, himself, added to this possibility by stating. In 1937, *he had been working on a superweapon from 1900 until the present time and did design and* _test_ *a charged-particle beam projector called Teleforce.*

In Tunguska we find tragedy, with no cause. I know some have indicated a massive meteor hit, but no remains suggest that at all. Some have stated a black hole caused it, but that would certainly have been more damaging to our planet. Some say it was an atomic event, but atomic weapons had not been created. What had been created in 1908 was Wardenclyffe Tower.

Here is what we do know. The Wardenclyffe tower was taken over by the military and blown up in 1917 and the infrastructure was so massive that their first attempts were useless against whatever was inside. We can believe our military had believed that Tesla's creation would be of upmost interest to our enemies for them to do something so visible.

The lingering question is how? Did he really have the power to blast away thousands of square kilometers of Russian wilderness? What we do know is this time is marked by his strange transformation into looking like the great science fiction actor, Peter Cushing. The image left is Tesla and the other two are Cushing.

  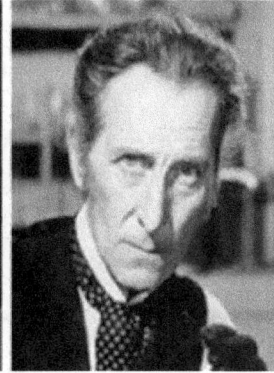

I don't know if this "Teleforce" event ever happened in such a dramatic way, but Nikola began to look at the power of the "self" right afterwards.

1909 Anthropics

In a New York Times Article, Tesla revealed the entire Theory of Anthropics decades before others. Tesla stated:

At mankind's command, almost without effort on his part, old words would vanish and new ones would spring into being. He could alter the size of the planet, control the seasons, adjust its distance to the sun; guide it on its eternal journey along any path he might choose, through the depths of the universe. He could make planets collide and produce his own suns and stars, his heat and light, he could originate life in all its infinite forms. To cause, at will, the birth and death of matter would be man's grandest deed, which would make him the mastery of physical creation, make him fulfil his ultimate destiny.

Remember this is way before Niels Bohr's quantum Mechanics science forced the concepts of multiverse generation of events established by cognizant observers and the whole science of Participatory Anthropics, but Nikola was capable of understanding it before anyone else. Later most would adopt the idea of multiverse reality where a cognizant viewer must witness an event or it does not happen. Today this seems fairly easy to understand and most know about the example of Schrodeger's cat being both dead and alive until someone actually witnesses the event. Years after Tesla's earth shattering claim, when Einstein was asked, *"If a tree falls in the woods and no one is there to hear it fall, does it still make a sound?"* Einstein simply said, *"There is no tree!"*

The following image is of Einstein and Tesla. While they would argue about things going faster than the speed of light, they were still friends.

When asked how it felt to be the smartest man alive, Albert Einstein replied---

"I don't know, you'll have to ask Nikola Tesla."

1910 Vibrational Matter

Tesla spelled out the idea of matter developing from Aether in 1908. While John Keely had initially expressed this concept, Einstein would not adopt it for many years after Tesla.

"According to an adopted theory, every ponderable atom is differentiated from a tenuous fluid, filling all space merely by spinning motion, as a whirl of water in a calm lake. <u>By being set in movement this fluid, the Aether, becomes gross matter</u>. Its movement arrested, the primary substance reverts to its normal state. It appears, then, possible for man through harnessed energy of the medium and suitable agencies for starting and stopping ether whirls to <u>cause matter to form and disappear</u>. At his command, almost <u>without effort</u> on his part, old worlds would vanish and new ones would spring into being. He could alter the size of this planet, control its seasons, adjust its distance from the sun, guide it on its eternal journey along any path he might choose, through the depths of the universe. He could make planets collide and produce his suns and stars, his heat and light; he could originate life in all its infinite forms. To cause at will the birth and death of matter would be man's grandest deed,

which would give him the mastery of physical creation, make him fulfill his ultimate destiny."

Predicted in 1916 by Albert Einstein to exist on the basis of his theory of general relativity, gravitational waves theoretically transport energy as gravitational radiation. In 1922 he said the following:

"Recapitulating, we may say that according to the general theory of relativity space is endowed with physical qualities; in this sense, therefore, there exists an Aether. According to the general theory of relativity space without Aether is unthinkable; for in such space there not only would be no propagation of light, but also no possibility of existence for standards of space and time, nor therefore any space-time intervals in the physical sense. But this ether may not be thought of as endowed with the quality characteristic of ponderable media, as consisting of parts which may be tracked through time. The idea of motion may not be applied to it"

When Nikola was 82, he reiterated his first words.

"The primary substance, thrown into infinitesimal whirls of prodigious velocity, becomes gross matter; the force subsiding, the motion ceases and matter disappears, reverting to the primary substance".

These gravity waves behave in similar ways to many other types of waves. Tesla's greatest inventions were all based on the study of waves. He always considered sound, light, heat, X-rays and radio waves to be related phenomena that could be studied using the same sort of math. For this reason, there exists the possibility that Tesla had extended this thinking to gravity. In 1932, he started the following:

114

"There is no more energy in matter than that received from the environment. We read a great deal about matter being changed into force and force being changed into matter by the cosmic rays. This is absurd. It is the same as saying that the body can be changed into the mind, and the mind into the body. We know that the mind i's a functioning of the body, and in the same manner force is a function of matter. Without the body there can be no mind, without matter there can be no force.

Vibrational model of matter had a problem

Sorry for this next section and I will be a quick as I can, but everything you thought you knew about life, matter and light are not exactly right. To really understand Electromagnetics of even what matter really is, we cannot use the old Length, width, and height definitions of existence. Long ago we found this concept simply doesn't work and Tesla knew this just like Einstein and all the rest. In Quantum Mechanics and the relativistic descriptions of vibrational matter, there is no "end" to anything. A piece of wood, for instance looks like it has a surface, but its vibrations actually continue outward all the way to the end of the universe and everything else has this same construct. Even a single atom vibrates outward from its center "called a vibrational standing wave" and extends outward in all directions as it goes outwards, its "Influence" is reduced quickly. What Tesla understood was increasing the vibrations of an electromagnetic wave made it affect the universe more.

Vibrational Matter and String Dimensions

We are only now beginning to understand the details of Tesla's vibrational matter model. As a component of this logic, it has been observed that there are potentially 12 different dimensional strings that all become existent when they vibrate. Today this is called "Quantum Fluctuations" but is the same thing. Increases in the rapidity of these vibrations increase the interaction of dimensional qualities in our universe. As an example; matter gets denser as its vibration increases and light or Electro-magnetism becomes more powerful as its vibrations increase.

Matter Dimensional Dynamo-Three mutually perpendicular vibrational components of matter are Aether [potential for matter], Gravity [Kinetic component of matter], and the combined Aether-Gravitational waves [that establish matter.].

Electro-magnetic Dimensional Dynamo-Like matter, all E-M or Light in the universe has a three-dimensional characteristic with Electricity [potential to do work], Magnetism [Kinetic component of Light], and Electro-magnetic waves [that establish Light]

Life and "Space-time" Dimensional Dynamos-All life in the universe has a three-dimensional characteristic we call the self, spirit, and soul. All space-time in the universe has a three-dimensional characteristic we call the Time, space, and space-time. These last 2 we don't need to get into to discuss Tesla. The following generally describes how all these "Dimensions" fit together to build out universe. Don't worry if this is just out there for you. It is only background to describe vibrational matter and vibrational essence of Electro-magnetism used by Tesla.

116

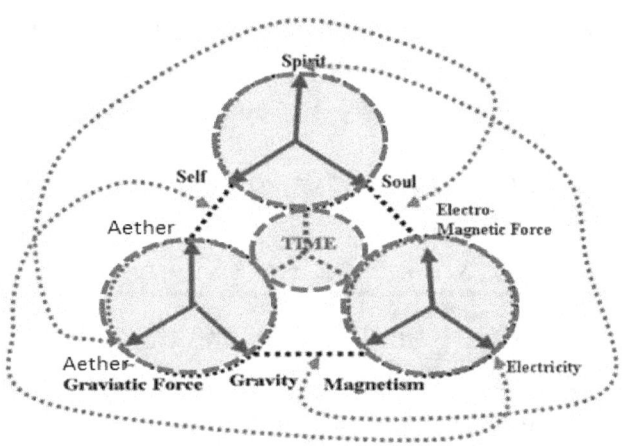

Symmetry Not Conservation

So, you are told about Conservation of Energy, Conservation of matter with the push towards Entropy, and some may even have told you about conservation of space-time and perhaps even conservation of life. All of these things seem to occur in our universe, but without an adjoined universe, just how is all this done? The answer comes from what scientists call super symmetry. This requires 2 LINKED universes.

As it turns out, symmetry is more important and more realistic than conservation of universal elements. You may know about the Theory of Conservation of Energy. Energy simply changes state rather than dissipates. It has an endless oscillation or vibration between Static and Kinetic elements. In a way, this is a true statement and certainly we can look at our universe in this way and things seem to fit together. What if I were to tell you that matter is a characteristic of the out-waves from our universe and Electro-magnetic energy is a characteristic of in-waves from an outside universe. You may begin to see that we don't conserve energy. It is continuously replenished. As our universe outputs out-waves, they are

117

converted to in-waves and returned to our universe. I know I'm losing you so I will be a brief as I can.

The universe dual MUST run symmetrically. The only way to apply force in this universe is from an outside force. AND Guess what!!! All dimensional qualities of this universe MUST BE symmetric with its universe dual.

Oh, you are a smart one!!! You are thinking, "That's not symmetry, the examples I keep bringing up show the opposite to symmetry.

Got You With This One!!!!

Please don't worry about this, but our linked universe operates with backward time. Therefore, decreasing matter backward in time is exactly expanding matter in their time. Both universes experience expanding masses as matter is created over here, anti-matter [defined as matter going backwards in time] is increased over there in backward time.

- If **kinetic energy** decreases, it increases in our joined universe which in-turn causes our STATIC Energy to Increase and vice-versa. [If we make a conversion of energy in our universe, the opposite WILL occur in the joined universe.]

- If **Gravitational energy** decreases, it increases in our joined universe which in-turn causes the exact opposite to increase in our universe. This increase makes us have an apparent reduction in mass so we will have to look at it a little. [If we force a reduction in gravity, our joined universe WILL experience an increase in magnetism to compensate.]

- If **Photonic [Electro-magnetic] energy** increases, it

118

increases in our joined universe which in-turn causes the exact opposite to decrease in our universe [Because in-waves turn into out-waves, electromagnetism become particles in the adjacent universe so mass would reduce.]

- If **Internal Life energy** decreases, it increases in our joined universe which in-turn causes our External Life Energy to Increase and vice-versa. [The main thing is that as internal life here increases, it WILL decrease over there.]

- As Time vector goes in one direction, it MUST go in the exact opposite direction on a joined universe. [Time must go backwards. This, of course is only from our point of view. If this were not so, time would eventually escape our universe and be lost]

Certainly, these may not be the exact duals of our universe, but hopefully, it will give you an idea about how each one of the elemental parts connect with both internal dual dimensions and external anti-dual dimensions.

To carry it one step further, we can establish that the energy bonds for the dimensional dynamo that develops matter are "Static" and "centripetal" forces. The energy bonds for the dimensional dynamo that produced light and Electro-magnetic waves are associated with "Kinetic" and "centrifugal" forces. The opposite would be true in the joined universe. The anti-matter dynamo would be associated with kinetic energy and the anti-photon Dynamo would be associated with Static energy in that world. All this stuff is good information, but Tesla was developing a way to expand our manipulation of elements in our universe, especially Magnetism, Gravity, Matter, and light. As sort of a cheat sheet, the characteristics of the elemental parts of our universe are described in the following chart. Don't worry about it too much it simply shows that everything can be reduced to a vibrational equation. Tesla was working on how to change the vibration.

Structural	Operational	Life	Time
$E_m= ½ Mz^2$	$E_e= ½ Cv^2$	$E_l= ½ Ay^2$	$E_t= ½ Rx^2$
E_m=Mass Potential	E_e= Electrical Potential	E_l=Life Potential	E_t=Time Potential
M=Mass capacity	C- Electrical capacity	A=Anthropic Capacity	R= Time capacity
z= Aetheric amplitude	V= Electrical Amplitude	y= Life amplitude	x= Space distance
$Z_m=(2πfM)^{-1}$	$Z_e=(2πfC)^{-1}$	$Z_l=(2πfA)^{-1}$	$Z_t=(2πfR)^{-1}$
Z= Aetheric Reactance	Z=Electrical Reactance	Z=Life Reactance	Z_t=Time Reactance
$E_g=½ Gv^2$	$E_m= ½ LI^2$	$E_s=½ SD^2$	$E_h=½ WG^2$
E=Kinetic Gravity	E=Kinetic Magnetism	E=Kinetic Soul	E=Heaven Energy
G=Induced Gravity	L=Magnetic Induction	S= Soul Induction	W= Heaven Induction
v= Gravity distance	I= Magnetic Control	D= Soul Control	G= Heaven Control
$Z_g=2πfG$	$Z_m=2πfL$	$Z_s=2πfS$	$Z_h=2πfM$
Z_g=Gravity Reactance	Z_m=Magnetic Reactance	Z_s=Soul Reactance	Z_h=Heaven Reactance
$E_n=MC^2$	$E_p=CC^2$	$E_l=AC^2$	$E_t=AC^3$
E=Mass Energy	Ep=Photonic Energy	Elf=Life-force Energy	Est=Space-time Energy
$R= GM^{-1/2}$	$R= LC^{-1/2}$	$R= AD^{-1/2}$	$R= AR^{-1/2}$
R=Mass Resonance	R=Photonic Resonance	R=Life Resonance	R=Time Resonance

The next graphic may put perspective on this whole vibrational world and how one thing can affect another. In the chart, Mass vibrations leave our universe and enter our linked universe just as their mass vibrations leave that universe.

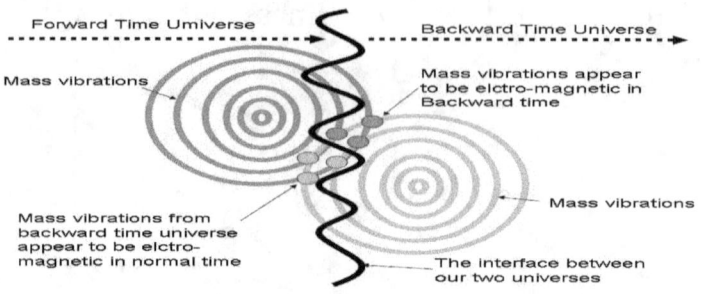

Forward Time Umiverse

Backward Time Universe

Mass vibrations

Mass vibrations appear to be elctro-magnetic in Backward time

Mass vibrations from backward time universe appear to be elctro-magnetic in normal time

Mass vibrations

The interface between our two universes

Life Light and Light are all the same thing. Change matter into something similar to light and gravity has no effect. In fact, one can even make matter invisible.

Mass Invisibility

Vibrational matter is only affected by gravity when it is visible to it. In fact, the only reason one cannot move through a "solid" wall is that its vibrations are in sync. By adding out of phase vibrational matter it does the same thing as noise cancelling headphones. By having 2 masses vibrating out of phase, they don't exist. The same happens with gravity and that is what Tesla was after, excite gravity until it disappears just like the ancient people did. The graphic shows a tams undulating [quantum fluctuating] but out of phase. The bottom line shows together they effectively have no vibration and disappear. Later we will see that John Hutchison has done this exact thing with high frequency ultra-high voltage signals. We can believe the same phenomenon was noticed by Tesla.

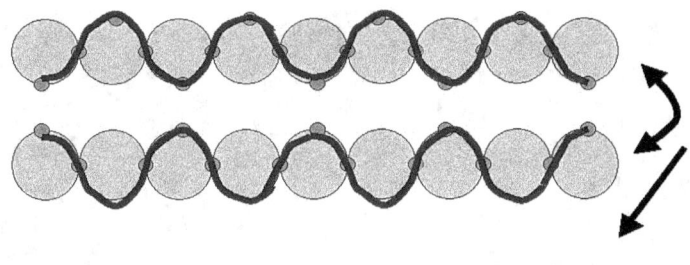

1910 Free Energy Machines

What if one could cool or heat a device at will using differential wind, temperature, or other characteristic that established difference in our ecosphere? Nikola came up with ways to use this "free energy".

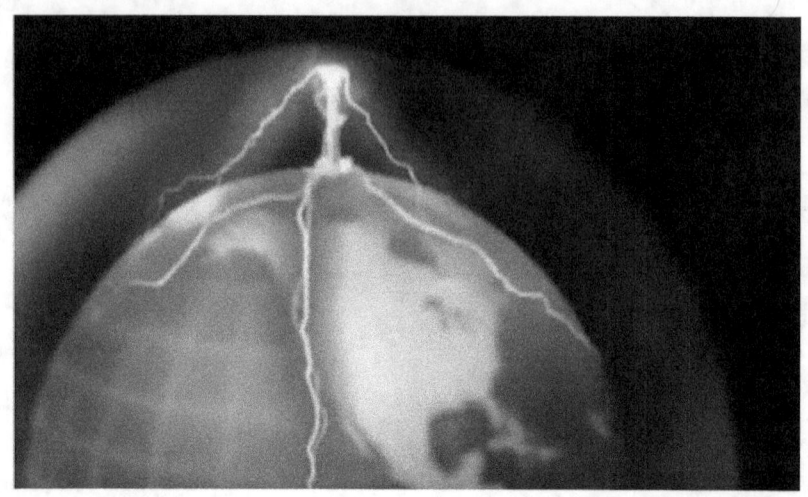

*"I read some statements from Carnot and Lord Kelvin which meant virtually that it is impossible for an inanimate mechanism or self-acting machine to cool a portion of the medium below the temperature of the surrounding, and operate by the heat abstracted. These statements interested me intensely. Evidently a living being could do this very thing, and since the experiences of my early life which I have related had convinced me that a living being is only an automaton, or, otherwise stated, a **"self-acting-engine,"** I came to the conclusion that it was possible to construct a machine which would do the same. As the first step toward this realization I conceived the following mechanism. Imagine a <u>thermopile consisting of a number of bars of metal extending from the earth to the outer space beyond the atmosphere. The heat from below, conducted upward along these metal bars, would cool the earth</u> or the sea or the air, according to the location of the lower parts of the bars, and the result, as is well known, would be an electric current circulating in these bars. <u>The two terminals of the thermopile could now be joined through an electric motor, and, theoretically, this motor would run on and on, until the</u>*

124

media below would be cooled down to the temperature of the outer space. This would be an inanimate engine which, to all evidence, would be cooling a portion of the medium below the temperature of the surrounding, and operating by the heat abstracted". "Much of this task on which I have labored so long remains to be done. A number of mechanical details are still to be perfected and some difficulties of a different nature to be mastered, and I cannot hope to produce a self-acting machine deriving energy from the ambient medium for a long time yet, even if all my expectations should materialize".

Strangle I have noticed new ideas using this same concept with an elevator of sorts between the ground and geostationary craft. Tesla was way ahead of his time in this regard. Another uses RF and photonic signal collection to power automobiles. We will look at that later, but while all this was going on, Nikola was changing from his Peter Cushing persona to another science fiction actor named Michael Rennie. His Peter Cushing days were out and these were considered, by some, to be Tesla's "Michael Rennie" years due to his uncanny resemblance to Klaatu from the movie *"The Day the Earth Stood Still"* [Klaatu is shown bottom row first. This is followed by "the Keeper" from the *"Lost in Space"* TV show in 1966 and the last is later as a detective who was an earthling all played by Michael Rennie/ Nikola look-alike.

Even though the movie would not debut until 1951, Tesla felt self-conscious about the resemblance and was assured that his middle hair part would make him stand out [See above third]. Michael Rennie on the bottom and there is absolutely no evidence Nikola Tesla every said the phrase *"Klaatu barada nikto"* to the robot Gort or anyone else, but he did begin to look more at aircraft and the possibility of refueling without fuel..

1910 E-M Fuel For Aircraft

Tesla had written about supplying energy to aircraft flying around so that they would need no internal fuel source. While this just sounded stupid to many, Artists began showing the Wardenclyffe tower powering all types of aircraft with some type of Electro-Magnetic charge of some kind as shown in the following collage.

Not only was the idea feasible and revolutionary, this same idea is believed to have been the number one-way ancient aircraft were fueled. The Indian historical references talk about Mercury as a critical part of the mechanism but there is not enough details for Nikola to have known about this. One can believe his unseen advisors aided his insights in this regard. Luckily, many, many images that depict this very thing can be found from ancient Sumeria, Babylonia, Assyria, Egypt. A few are shown in the following collage.,

Forget the Sumerian and Babylonian operators of the tesla coil type devices are wearing fish outfits. Almost all the depictions of this type of radiating device shows a vehicle flying in the air over it. Some show lightning bolt being sent to the flying machines. The images below are from Egypt and Assyria. While the fish outfits changed, the Tesla looking device did not and the aircraft are still getting fueled.

The question might be, how did Nikola know? Had he seen these artifacts or was something else going on? He decided wings got in the way of flying.

1911 Wingless Aircraft

This could have been his greatest achievement and we know little about While he indicated he had pondered methods for flying early in his life, Nikola decided to concentrate on his fascination diligently in 1911. During this time. he described an onboard electro-propulsive means to be used on his "ideal" flying machine, either manned or unmanned, controlled "mechanically" by a pilot on board, or "remotely by wireless energy" by a controller from the ground. It did not refer to an "airplane" since his machine was to have no wings, ailerons, propellers, or outer appendages of any kind.

New Propulsion System

"Propulsion here means an on-board system for perpetuation of motion, by electromagnetic means, supplied with electrical power by either on-board generator, or by electrical energy and control signals transmitted by power beam from the ground." Besides "in-atmosphere" travel, the idea of interplanetary travel appealed to Tesla, with the idea that, so long as the electrical energy for propulsion could be transmitted from the earth, a space ship would require no fuel tanks. In a letter to his friend and financial supporter B.A. Behrend in the 1930's, Tesla referred cryptically to his electro-propulsive discovery: *"What I shall accomplish by*

that other invention I came specially to see you about, I do not dare to tell you. This is stated in all seriousness."

Electro-Levitation

Tesla had realized that all solid bodies contain an "electrical content", and that these circuits behave as resonant cavities, which interact electromagnetically with rapidly varying electrostatic forces and ether to determine their gravitational interactions and movements in space. While new to Tesla, apparently, ancient civilizations had mastered the whole levitated flying saucer thousands of years ago. The first set of examples are from ancient Egypt and India.

Next, we have more examples from Sumeria, Mexico, Guatemala, and China, as the design and operation of a wingless aircraft was common throughout the world 5-thousand years ago.

Another thing that seems pretty probable is they went outside our atmosphere and had to wear some type of breathing equipment as shown in the small sample following.

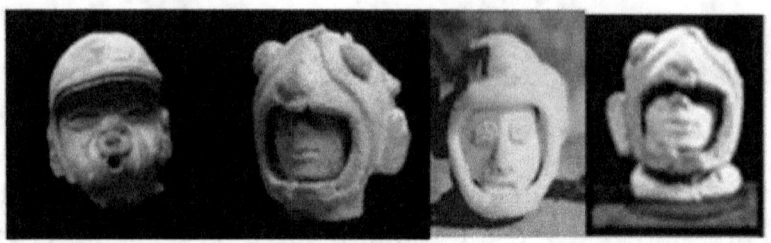

These ancient aircraft apparently are still being produced as sighting almost every day for thousands of years has shown us its common use. Here are just a few of the sightings.

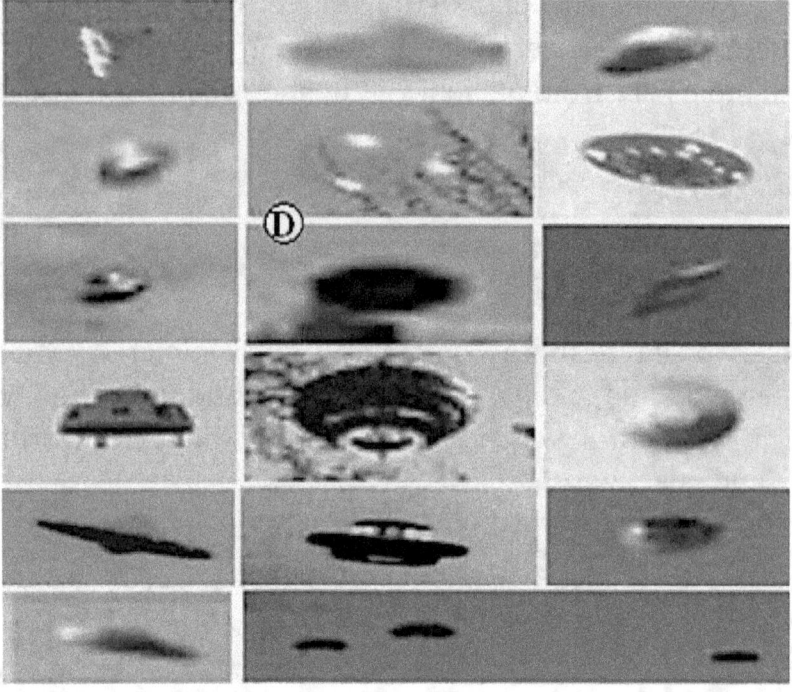

It is believed these theories were tested and confirmed to some degree in his Colorado Springs experiments, but the US Government was not interested. Somehow, this information became known to the Germans around 1937 so that they began a massive build-up of flying saucer prototypes.

Several different designs were built and flown during World War II as shown next.

Some were dropped from airplanes and they were seen straffing American ships and shown in the video stills.

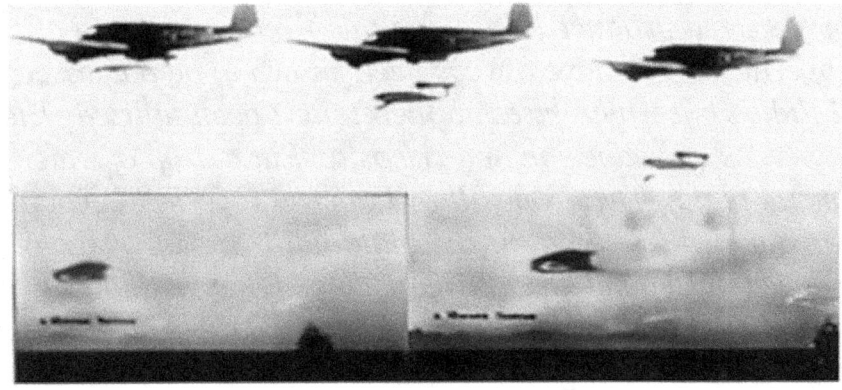

Wingless

Let's see what Nikola actually said about his discovery. Nikola Tesla told *The New York Herald* that he was working on an anti-gravity *"flying machine"*. *"My flying machine will have neither wings nor propellers. You might see it on the ground and you would never guess that it was a flying machine. Yet it will be able to move at will through the air in any direction with perfect safety, at higher speeds than have yet been reached, regardless of weather and oblivious of "holes in the air" or downward currents. It will ascend in such currents if desired. It can remain absolutely stationary in the air, even in a wind, for great length of time. Its lifting*

133

power will not depend upon any such delicate devices as the bird has to employ, but upon positive mechanical action."

Super Light Engine

Later, he said the following: *"I have accomplished what mechanical engineers have been dreaming about ever since the invention of steam power, that is the perfect rotary engine. It happens that I have also produced an engine which will give at least twenty-five times as much power to a pound of weight as the lightest weight engine of any kind that has yet been produced. In doing this I have made use of two properties which have always been known to be possessed by all fluids, but which have not heretofore been utilized. These properties are adhesion and viscosity. Put a drop of water on a metal plate. The drop will roll off, but a certain amount of the water will remain on the plate until it evaporates or is removed by some absorptive means. The metal does not absorb any of the water, but the water adheres to it. The drop of water may change its shape, but until its particles are separated by some external power it remains intact. This tendency of all fluids to resist molecular separation is viscosity. It is especially noticeable in the heavier oils. It is these properties of adhesion and viscosity that cause the "skin friction" that impedes a ship in its progress through the water or an aeroplane in going through the air. All fluids have these qualities--and you must keep in mind that air is a fluid, all gases are fluids, steam is fluid. Every known means of transmitting or developing mechanical power is through a fluid medium.'*

Vaneless Turbine

Now, suppose we make this metal plate that I have spoken of circular in shape and mount it at its center on a shaft so that

134

it can be revolved. Apply power to rotate the shaft and what happens? Why, whatever fluid the disk happens to be revolving in is agitated and dragged along in the direction of rotation, because the fluid tends to adhere to the disk and the viscosity causes the motion given to the adhering particles of the fluid to be transmitted to the whole mass. Here, I can show you better than tell you."

Demonstration of Aircraft Turbine

Dr. Tesla led the way into an adjoining room. On a desk was a small electric motor and mounted on the shaft were half a dozen flat disks, separated by perhaps a sixteenth of an inch from one another, each disk being less than that in thickness. He turned a switch and the motor began to buzz. A wave of cool air was immediately felt. There we have a disk, or rather a series of disks, revolving in a fluid--the air. You need no proof to tell you that the air is being agitated and propelled violently. If you will hold your hand over the center of these disks--you see the centers have been cut away--you will feel the suction as air is drawn in to be expelled from the peripheries of the disks. Now, suppose these revolving disks were enclosed in an air tight case, so constructed that the air could enter only at one point and be expelled only at another. It is an engine that does all that engineers have ever dreamed of an engine doing, and more. Down at the Waterside power station of the New York Edison Company, through their courtesy, I have had a number of such engines in operation. In one of them the disks are only nine inches in diameter and the whole working part is two inches thick. With steam as the propulsive fluid it develops 110-horse power, and could do twice as much. You have got what Professor Langley was trying to evolve for his flying machine--an engine that will

135

give a horse power for a pound of weight, With a thousand horsepower engine, weighing only one hundred pounds, imagine the possibilities in automobiles, locomotives and steamships-- And it makes the flying machine practical.

Tesla's Flying Machine Description

Now you have struck the point in which I am most deeply interested--the object toward which I have been devoting my energies for more than twenty years--the dream of my life. It was in seeking the means of making the perfect flying machine that I developed this engine---my flying machine-- will be heavier than air, but it will not be an aeroplane. It will have no wings. It will be substantial, solid, stable. You cannot have a stable airplane. The gyroscope can never be successfully applied to the airplane, for it would give a stability that would result in the machine being torn to pieces by the wind, just as the unprotected aeroplane on the ground is torn to pieces by a high wind. My flying machine will have neither wings nor propellers. You might see it on the ground and you would never guess that it was a flying machine. Yet it will be able to move at will through the air in any direction with perfect safety, higher speeds than have yet been reached, regardless of weather and oblivious of "holes in the air" or downward currents. It will ascend in such currents if desired. It can remain absolutely stationary in the air, even in a wind, for great length of time. Its lifting power will not depend upon any such delicate devices as the bird has to employ, but upon positive mechanical action." The original shape of this new vehicle type is shown below.

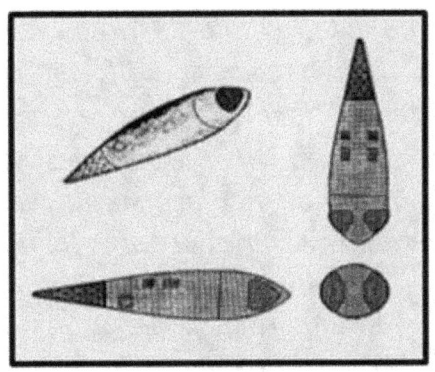

Stabilization of Aircraft

Through gyroscopic action of my engine it will get stability, assisted by some devices I am not yet prepared to talk about," he replied. --All I have to say on that point is that my airship will have neither gas bag, wings nor propellers. It is the child of my dreams, the product of years of intense and painful toil and research. I am not going to talk about it any further. But whatever my airship may be, here at least is an engine that will do things that no other engine ever has done, and that is something tangible."

Cosmic Ray Engine

On his 76th birthday Tesla said, *"I have succeeded in Harnessing 'Penetrating Rays' to Operate Small Motive Device. I have harnessed the cosmic rays and caused them to operate a motive device. Cosmic ray investigation is a subject that is very close to me. I was the first to discover these rays and I naturally feel toward them as I would toward my own flesh and blood. I have advanced a theory of the cosmic rays and at every step of my investigations I have found it completely justified. The power output was many thousand times that of a Crookes' radiometer. The attractive features of the Cosmic rays are their constancies. They shower down on us throughout the whole 24 hours, and if a plant is developed to use their power it will not require devices for storing energy as would be necessary with devices using wind, tide or sunlight. All of my investigations seem to point to the conclusion that they are small particles, each carrying so small a charge that we are justified in calling them neutrons. They move with great velocity. More than 25 years ago I began my efforts to harness the cosmic rays and I can now state that I have succeeded in operating a motive device by means of them.* The cosmic ray ionizes the air, setting free many charges—ions and electrons. These charges are captured in a condenser which is made to discharge through the circuit of the motor. I have hopes of building my motor on a large scale, but circumstances have not been favorable to carrying out my plan."

Standing Wave Levitation

The power would be transmitted by creating "standing waves" in the earth by charging the earth with a giant electrical oscillator that would make the earth vibrate electrically in the same way a bell vibrates mechanically when it is struck with a hammer. --I use the conductivity of the earth itself, and in this I need no wires to send electrical energy to any part of the globe."

Round Ship

The hull is best made double, of thin, machinable, slightly flexible ceramic. This becomes a good electrical insulator, has no fire danger, resists any damaging effects of severe heat and cold, and has the hardness of armor, besides being easy for magnetic fields to pass through. The inner hull is covered on its outside by wedge shaped thin metal sheets of copper or aluminum, bonded to the ceramic. Each sheet is 3

to 4 feet wide at the horizontal rim of the hull and tapers to a few inches wide at the top of the hull for the top set of metal sheets, or at the bottom for the bottom set of sheets. Each sheet is separated on either side from the next sheet by 1 or 2 inches of uncovered ceramic hull. The top set of sheets and bottom set of sheets are separated by about 6 inches of uncovered ceramic hull around the horizontal rim of the hull. The outer hull protects these sheets from being short-circuited by wind-blown metal foil, heavy rain or concentrations of gasoline or kerosene fumes. If unshielded, fuel fumes could be electrostatically attracted to the hull sheets, burn and form carbon deposits across the insulating gaps between the sheets, causing a short-circuit. In space, the outer hull with a slight negative charge, would absorb hits from micro-meteorites and cosmic rays. Any danger of this type that doesn't already have a negative electric charge would get a negative charge in hitting the outer hull, and be repelled by the metal sheets before it could hit the inner hull. This wouldn't work well on a very big meteor, I might add. The hull can be made in a variety of shapes; sphere, football, disc, or streamlined rectangle or triangle, as long as these metal sheets, "are of considerable area and arranged along ideal enveloping surfaces of <u>very large radii of curvature</u>,"
A couple of artist conceptions of his design are shown below.

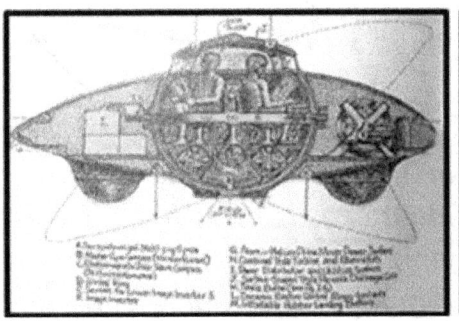 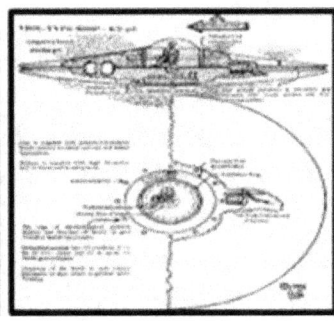

Early Desire for Levitation

Tesla gave us information that he had worked on the flying machine in his head for many years. Here is what he said.

In the second year at that institution I became obsessed with the idea of producing continuous motion thru steady air pressure. The pump incident, of which I have told, had set afire my youthful imagination and imprest me with the boundless abilities of a vacuum. I grew frantic in my desire to harness this inexhaustible energy but for a long time I was groping in the dark. Finally, however, my endeavors crystallized in an invention which was to enable me to achieve what no other mortal ever attempted. Imagine a cylinder freely rotatable on two bearings and partly surrounded by a rectangular trough which fits it perfectly. The open side of the trough is closed by a partition so that the cylindrical segment within the enclosure divides the latter into two compartments entirely separated from each other by air-tight sliding joints. One of these compartments being sealed and once for all exhausted, the other remaining open, a perpetual rotation of the cylinder would result, at least, I thought so. A wooden model was constructed and fitted with infinite care and when I applied the pump on one side and actually observed that there was a tendency to turning, I was delirious with joy. Mechanical flight was the one thing I wanted to accomplish although still under the discouraging recollection of a bad fall I sustained by jumping with an umbrella from the top of a building. Every day I used to transport myself thru the air to distant regions but could not understand just how I managed to do it. Now I had something concrete—a flying machine with nothing more than a rotating shaft, flapping wings, and—a vacuum of unlimited power! From that time on I made my daily aerial excursions in a vehicle of comfort and luxury as might have befitted King Solomon. It took years before I understood that the <u>atmospheric pressure acted at right angles to the surface of the cylinder and that the slight rotary effort I observed was due to a leak.</u> Though this knowledge came gradually it gave me a painful shock.

Memory of Early Design

I explained a flying machine I had conceived, not as an illusionary invention, but one based on sound, scientific principles, which has become <u>realizable thru my turbine</u> and will soon be given to the world. I had written, and a package of calculations relating to solutions of an unsolvable integral and to my flying machine. As stated on a previous occasion, when I was a student at college I conceived a flying machine quite unlike the present ones. The underlying principle was sound but could not be carried into practice for want of <u>a prime-mover of sufficiently great activity</u>. In recent years, I have <u>successfully solved this problem</u> and am now planning aerial machines devoid of sustaining planes, ailerons, propellers and other external attachments, which will be capable of immense speeds and are very likely to furnish powerful arguments for peace in the near future. Such a machine, sustained and <u>propelled entirely by reaction</u>, is shown on page 108 <u>and is supposed to be controlled either mechanically or by wireless energy.</u> By installing proper plants it will be practicable to project a missile of this kind into the air and drop it almost on the very spot designated, which may be thousands of miles away.

Stationary Flight- *It also would be able to <u>remain absolutely stationary in the air</u>. He described the engine as*

141

having 1000 horse power, and weighing only 100 lb. To make a comparison, one of the most powerful turbine engine of the modern commercial aviation, a single GE90-115B puts out over 110,000 horsepower (the double of Titanic's engines), weighing 18.260 lb. in empty condition.

Aerial Navigation-Dr. Tesla considered for a moment or two and then replied with great deliberation: *"The application of this principle will give the world a flying machine unlike anything that has ever been suggested before. It will have no planes, no screw propellers or devices of any kind hitherto used. It will be small and compact, excessively swift, and, above all, perfectly safe in the greatest storm. It can be built of any size and <u>can carry any weight</u> that may be desired"*

New York Times 1911

Tesla's New Engine - *"Mr. Charles Wilson Price, editor of the Electrical Review, does not say that Nikola Tesla's latest invention of a rotary engine operated by steam or gases will save most of the 30,000,000 horse power wasted annually by manufacturing plants in this country, power that is worth upward half a billion dollars. He has not said, as Mr. Tesla is reported saying, that the new engine will give the world a flying machine that, without planes or screw propellers, 'excessively swift, and, above all, perfectly safe in the greatest storm, will be small and compact compared with present machines, but capable of carrying shiploads of passengers."*

"The Tesla Turbine" - 1912

"I was a mechanical engineer before I was an electrical engineer, and besides, this principle was worked on in the

142

course of my search for the ideal motor for airships, to be used in conjunction with my invention for the wireless transmission of electrical power. For twenty years I worked on the problem, but I have not given up. When my plan is perfected the present-day aeroplanes and dirigible balloons will disappear, and the dangerous sport of aviation, as we know it now with its hundreds of accidents, and its picturesque birdmen, will give way to safe, <u>seaworthy airships, without wings</u> or gas bags, but supported and driven by mechanical means".

EM Powered- *"As I told you before when we were talking of the wireless transmission of power, the mechanism will be a development of the principle on which my turbine is constructed".*

Magnetic Levitation- *"It will be so tremendously powerful that it will make a <u>veritable rope of air above the great machine to hold it at any altitude</u> the navigators may choose, and also a rope of air in front or in the rear to send it forward or backward at almost any speed desired. When that day comes, airship travel will be as safe and prosaic as travel by railroad train today, and not much very different, except that there will be no dirt, and it will be much faster. One will be able to dine in New York, retire in an aero Pullman berth in a closed and perfectly furnished car, and arise to breakfast in London."*

Danger of Airplanes- *"The aeroplane is fatally defective. It is merely a toy—a sporting play-thing. It can never become commercially practical. It has fatal defects. One is the fact that <u>when it encounters a downward current of air</u> it is helpless. The —hole in the air— of which aviators speak is simply a downward current, and unless the aeroplane is high*

143

enough above the earth to move laterally but can do nothing but fall".

Great High Stability-*"The flying machine of the future—my flying machine—will be heavier than air, but it will not be an aeroplane. It will have no wings. It will be substantial, solid, stable. You cannot have a stable airplane. The gyroscope can never be successfully applied to the airplane, for it would give a stability that would result in the machine being torn to pieces by the wind, just as the unprotected aeroplane on the ground is torn to pieces by a high wind".*

More Magnetic Levitation-"You might see it on the ground and you would never guess that it was a flying machine. Yet it will be able to move at will through the air in any direction with perfect safety, higher speeds than have yet been reached, regardless of weather and oblivious of —holes in the air— or downward currents. It will ascend in such currents if desired. It can remain absolutely stationary in the air, even in a wind, for great length of time. Its lifting power will not depend upon any such delicate devices as the bird has to employ, but upon positive mechanical action".

Counter Rotating Turbine-"All I have to say on that point is that my airship will have neither gas bag, wings nor propellers, he said. It is the child of my dreams, the product of years of intense and painful toil and research. I am not going to talk about it any further. But whatever my airship may be, here at least is an engine that will do things that no other engine ever has done, and that is something tangible". US1,655,114 patented the stability of dual counter rotating turbines to power the flying machine or spaceship. Tesla Towers were to power saucers remotely if you understand sound waves in the Aether are Tesla's method of transmitted power. As described in some of the excepts have shown, Tesla believed aircraft could float rather than fly by levitation, but levitation has been in use for thousands of years.

Ancient Sonic Levitation

Let me just say this right now, Nikola did not invent levitation by means of sound. We can find dozens of documents that detail characteristics of this ancient method. The reason I am bringing this up is we find that a number of experiments done by Tesla used Mercury and possibly he made the claim from observations during his experiments. Here is a snippet of what he said.

"As stated on a previous occasion, when I was a student at college I conceived a flying machine quite unlike the present ones. The underlying principle was sound but could not be carried into practice for want of a prime-mover of sufficiently great activity. In recent years I have successfully solved this problem and am now planning aerial machines devoid of sustaining planes, ailerons, propellers, and other external attachments which will be capable of immense speeds and are very likely to furnish powerful arguments for peace in the future

Sonic Levitation

As I suggested, one method to move blocks into place or take them to the peak of a massive monument is to levitate them so let's see if there is confirmation Titans used to be able to move objects through the air.

Peruvian Sonic Levitation-The Aymara wrote that huge blocks were levitated and moved. *"The stones were miraculously lifted off the ground and transported by the sound of a trumpet."*

Levitating Mayans-In Mayan history we find more evidence. According to their history, sound was a major initiator of a levitation event. *"Workers, during ancient times, whistled a special tone to set huge boulders into place."*

Egyptian Levitation- Here we find details of moving huge masses by the ancient ones. *"A stone was wrapped in papyrus. Then a magician struck the wrapped stone with a rod. Once it was struck, some kind of sound began to emanate from the rod. Thus, the stone became completely weightless and moved through the air for about 50 meters. The process was repeated as required until the stone was in place."*

Levitating Phoenicians- According to an ancient Phoenician document, the following was found. *The father of Cronus devised Baetulia, contriving stones that moved as with having life.*

Indian Levitation- According to the 2000-year-old "Siddhanta" we find the following: *The magicians could become heavy or light at will.*

Bible Levitation- For this we find the Hebrew word Nasiy' which means "ruler" in 130 places in the Bible and was retranslated in 3 passaged mysteriously to say "vapor". Let's

147

re-read Psalm 135:7, Jeremiah 10:13, and Jeremiah 51:16. They all three say the same thing.

Jeremiah 10:13 When he uttereth his voice, there is a multitude of waters [planets] in the heavens, and <u>he causeth the Nasiy' [RULERS] to ascend</u> from the ends of the earth.

What about Crystalline Mercury?

As an Electronic Engineer by trade, I can tell you the normal frequencies that are output from Quartz Crystals don't appear to lift things off the ground directly. We can assume a different type of crystal was used in many of these ceremonies. That brings me to cinnabar. All through ancient Indian Text, they talk about cinnabar, Crystalline Mercury, as a major portion of the power sources they used for their flying vimanas.

In the description of making a flying ship engine, the following was required. *You must place a mixture of lode-stone, mercury, mica, and serpent-slough on the north and crystals in the center of the engine.* [If we just had serpent slough, we would be ready to design a flying ship engine. No one really knows what the words that were interpreted as Serpent slough really meant, but the crystals being used makes perfect sense as the vibrational tone of a crystal can change the materials around it as will be discussed later.]

Another passage says that *you must place one type of "mani" in sulfuric acid and another type place with magnetite, mica, mercury and, [of course], <u>serpent-slough</u>. All five crystals should be equipped with wires passing through glass tubes. Wires should be placed from the center in all directions, then a triple wheel will set the revolving motion and the two glass balls inside will turn and increase*

148

speed, rubbing each other. The resulting friction generates 100-degree power. [Well it's pretty clear that they are really talking about cinnabar here which is a crystalline form of mercury with sulfur.]

Still another passage indicates that *you must place the mercury engine with its iron heating apparatus below. By means of the power latent in the mercury which sets the driving whirlwind in motion, a man sitting inside may travel great distance in a marvelous manner.*

Another section talks about an engine with –*"Four strong mercury containers that must be built in the interior structure, heated by the controlled fire".*

Still more details are provided. It stated that *the vimanas [Flying ships] develop "thunder power" from the mercury.*

In the "Vaimanika Shastra" it talks about *a sort-of image receiver that used mercury and indicated that in the center, you must erect a 6-inch pivot and 4 tubes, made of vishvodara metal* [whatever that means].

Many other items are discussed and then "Vaimanika Shastra" says, *A vessel of mercury shall be placed at the bottom. In it, a crystal bead with a hole shall be placed. Through the hole should go wires, 8 parts of sun-power, and 12 parts of electric power. These should pass through the crystal into the mercury.... This enables the pilot to realize conditions of a certain area or inflict damage on the enemy.*

More Ancient Mercury

India wasn't the only place using or interested in Mercury. We can find evidence around the world. Here are a few finds, which suggest its worldwide appeal.

149

Greek Mercury-In ancient Greece, mercury was considered *the metal of the gods*. Hermes. [known as Mercury by the Romans]

Egyptian Mercury-Thoth was synonymous with the Geek's god "Mercury". Thoth was the magician ruler who possibly built the pyramids. In the writings of Thoth, there are claims that *he had a fantastic ship buried under the ground near the Sphinx*. The "ship" in the writings sounds like a massive flying vehicle so he probably used Mercury.

Russian Mercury-Russian investigators found what are believed to be parts of navigating equipment. These items are not large nor do they make up much of a machine. They are hemispherical, glass or porcelain, and one piece ended in a cone with *a drop of mercury inside*. The characteristics of these components seemed to be described in early Indian writings about navigation equipment for flying ships. The devices are estimated to be well over 8,000 years old so this mercury stuff has been used for machinery of some kind for a very long time.

Ancient Mining- At several locations around the world, Cinnabar [the stable crystalline structure containing mercury] was known to be mined at least 2500 years ago, so it must have been used for something. My guess is flying machines. What is yours?

Mercury was somehow used in ancient aitcraft to aid in their levitation into the air, but we just don't know how it was all done. Possibly Tesla was getting close to providing us the breakthrough we need for more reasonable air travel and he also described how we can harness the weather.

1915 Weather Control

In 1915, Nikola described another characteristic of his tower should it ever be completed- Weather control by EM transmission into the Ionosphere. He said *"The time is very near when we shall have the precipitation of the moisture of the atmosphere under complete control, and then it will be possible to draw unlimited quantities of water from the oceans, develop any desired amount of energy, and completely transform the globe by irrigation and intensive farming. A greater achievement of man through the medium of electricity can hardly be imagined."*

The PreMaya people wrote that the people before the great change 5000 years ago knew how to control the weather may give us insight about how he might have understood this thing, but what we really need to look at is how weather is made using Jets to steer the atmosphere. The image below left shows the normal airflow patterns around the world controlled by Polar and subtropical Jets.

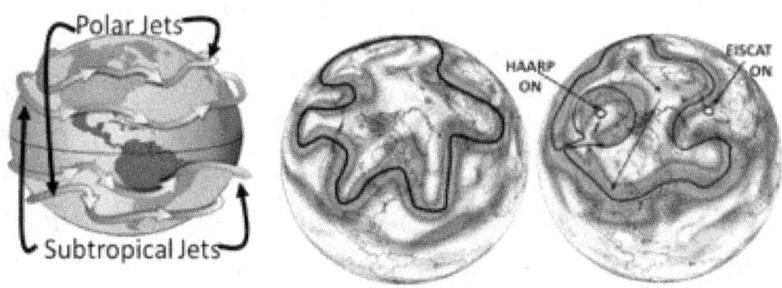

Now for something interesting. From the Climate Change Institute, we can look at the wind pattern from May 14 [middle] and November 14 2014 [right] and see something peculiar around two places titled HAARP and EISCAT. Essentially, HAARP is the United States Ionosphere Heater facility and the EISCAT is the equivalent European Ionosphere heater. This heating is accomplished by sending high frequency RF signals into the Ionosphere just like Nikola was doing. Let me show this again. The January 2014 weather map from NOAA shows a noticeable anomaly around the EISCAT Ionosphere Heater while in January 2015 below it we see it is not transmitting. We can believe Tesla's descriptions are being experimented with today.

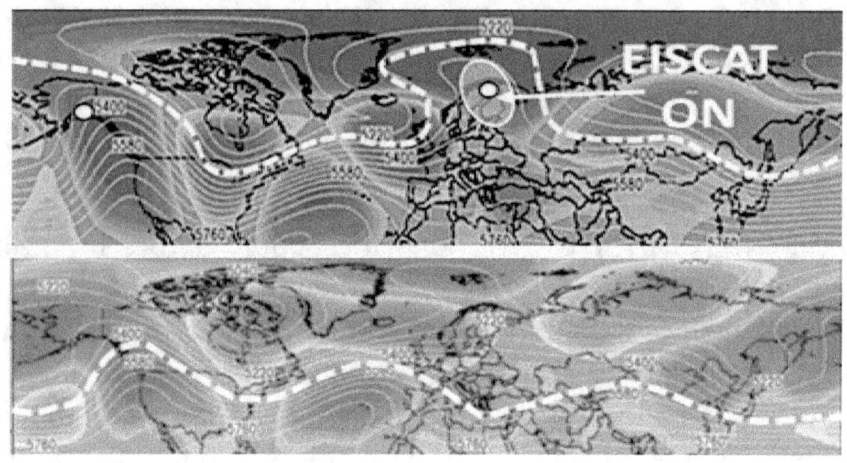

Just to make sure, let's see what the people see when the heating/weather control is attempted. On February 1996 the European Space Agency calmed viewers by explaining that *the DASI Digital All-sky imager of the EISCAT showed and auroral arc located slightly south of the facility during 2 ionosphere heating cycles. A development of spiral like forms occurred after the EISCAT was turned on.* Not only could this thing make its own Auroral Borealis like the HAARP system in Alaska, we can certainly believe it disrupted weather patterns. Its power is on display in the next images.

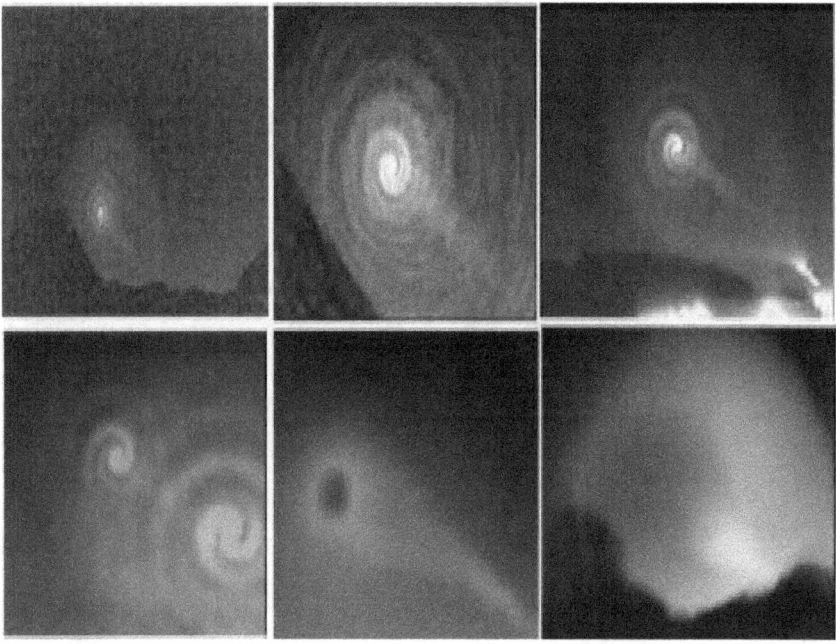

I might as well tell you that these Ionosphere heating stations not only modify weather patterns, they also are being used for around the world communication by sending RF signals into the ionosphere. Great success in communicating with submarines half way around the earth using this technique that Nikola had been experimenting with a hundred years ago. Later we will talk about missing documents from his files after death, but right now understand that Nikola was keenly aware of weather changes by his machine and he may even have seen the same cavitation and plasma patterns seen today when these things are turned on.

In 1999 EU Committee on Foreign Affairs Report- stated: *"Despite the existing conventions, military research* *is going on environmental manipulation as a weapon, as demonstrated for example by the Alaska-based HAARP system."*

Don't go thinking the US Government stole some of Nikola's notes to keep others from sending high frequency, high voltage messages into the Ionosphere, but when Nikola's tower was brought down for not good reason in 1917, one must wonder. Was is the Death ray or was is using weather as a weapon they were afraid of? Maybe they were starting to believe he was not a crackpot, but when he went to them about underwater radar, they pushed him aside again.

1915: Radar

In this case instead of telling where weather was bad, the Americans needed to know where submarines were that were killing our soldiers. Nikola Tesla, actually proposed multiple electrical schemes for locating submerged submarines. The reflected electric ray method is illustrated below and used his Radio system already tested and demonstrated. He also described a high-frequency invisible electric ray that, when reflected by a submarine hull, would cause phosphorescent screens on another or even the same ship to glow, giving warning that the U-boats are near. This would give a longer range that the radio system but was untested.

"By standing electromagnetic waves use we may produce at will, from a sending station, an electrical effect in any particular region of the globe; with which we may determine the relative position or course of a moving object, such as a vessel at sea, the distance traversed by the same, or its speed."

It looks like his old nemesis may have derailed this potential game changer against World War. The sad thing is that if Edison hadn't claimed one of Tesla's most crucial radio wave-based innovations to be "impractical" during World War I, countless lives could have been saved for having the advantage of being able to detect enemy submarines. Of course, it would be actualized until decades later. But just to think of what damage Edison's meddling had cost Nikola and the United States is sad indeed. The image below shows Nikola tuning one of his receivers, possibly one that could have detected submarine magnetic disturbances.

1915: Nobel Peace Prize

In 1915, a *New York Times* article announced that Tesla and Edison were to share the Nobel Prize for physics. Oddly, neither man received the prize, the reason being unclear. It was rumored that Tesla refused the prize because he would not share with Edison, and because Marconi had already received the one he deserved.

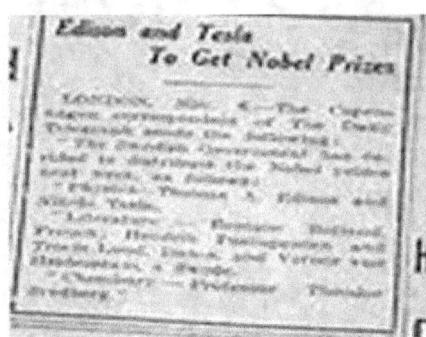

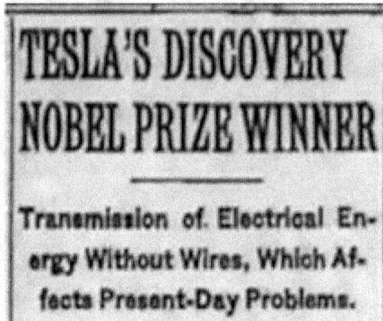

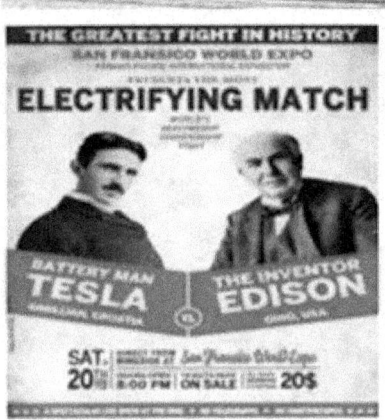

1926 Cell Phones

Speaking of applications of radios and wireless telephones, in 1902, Nathan Stubblefield was the first to develop "inductive linked" wireless ship to shore and telephone capability. The following image shows a demonstration of Nikola Tesla and Nathan Stubblefield's operational inductive cell phone. The box is around Tesla while Nathan is shown on the right end.

NBS Wireless Telephone Demonstration
Philadelphia - Belmont Park - 1902
1. A. Frederick Collins; 20. Nikola Tesla; 28. Nathan B. Stubblefield - Photo: NBS100.com

Nathan Stubblefield- In 1892, Nathan demonstrated an inductively linked radio communication device and later modified the design to allow ship to shore communication along a short span of a river. To do this he built a massive inductive loop as shown next. So long as someone stayed near a portion of the induction cable, it would receive signals like a large transformer.

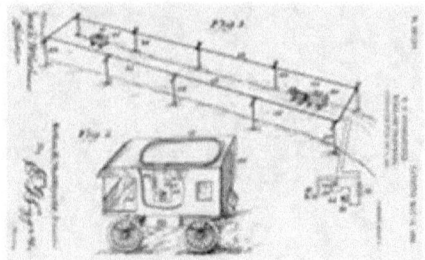 

His inductive radio and telephone are shown below. These would be the inspiration for Tesla's description of true cell phone operation. Stubblefield was never given credit so he locked himself in his barn and starved himself to death according to one story.

Here is what Nikola explained in 1926-*When wireless is perfectly applied, the whole earth will be converted into a huge brain, which in fact it is, all things being particles of a real and Rhythmic whole. We shall be able to communicated with one another instantly, irrespective of distance. Not only this, but through television and telephony we shall see and hear one another as perfectly as though we were face to face, despite intervening distances of thousands of miles, and the*

instruments through which we shall be able to do all this, will fit into our vest pocket.

Just to let you know how great television was when he said all this, by the end of 1926 a Japanese researcher Kenjiro Takayanagi demonstrates a system that used a mechanical disk in a transmitter and a cathode ray tube in the receiving device. With it he transmits 40 still images of a Japanese character, so his idea was far exceeding what they called TV and the telephone wasn't much better and it required cables between people. The TV transmitter is shown next. And the telephone [telephone poles not shown]. These could not fit in a pocket.

1926 Baird "Falkirk" Transmitter

1927: Vertical Take-off/Land Airships

Drone Aircraft-Tesla proposed that electrically-powered airships would transport passengers from New York to London in three hours, traveling eight miles high to a <u>preselected destination or for a remote aerial strike</u>. He was never given credit for his invention. However, today, we have unmanned drones carrying out combat missions, supersonic airplanes that fly at amazing speeds and space shuttle technology that can circle miles above the ground.

Magnetically powered with no internal fuel-He also imagined that airships might draw their power from the very atmosphere, never needing to stop for refueling. Unmanned airships might even be used to transport passengers the Earth in the upper atmosphere

Vertical take-off and Land vehicles- His patent shown next included characteristics of both helicopters and vertical take of airplanes by switching drive methods in the air.

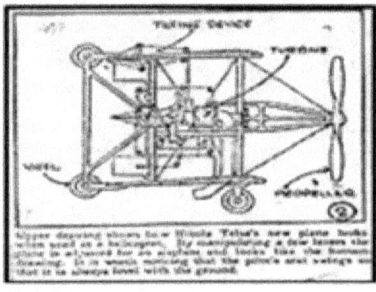 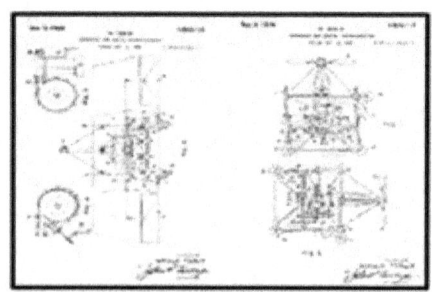

Next we see a modeled device built around the patened elements. The first image shows the take off position and, after transition, the second image shows the forward drive mode of operation, similar to some of our most exotic jets.

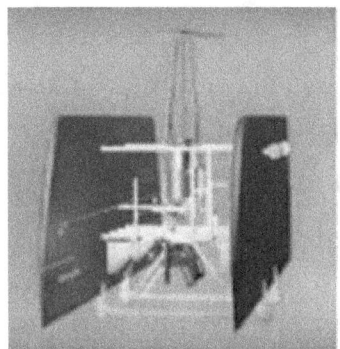 

1932: Cosmic Drive

On July 10, 1932, reporter, John J. A. O'Neill witnessed and wrote about the following. Here is what Nikola had to say.

I have harnessed the cosmic rays and caused them to operate a motive device. Cosmic ray investigation is a subject that is very close to me. I was the first to discover these rays and I naturally feel toward them as I would toward my own flesh and blood. I have advanced a theory of the cosmic rays and at every step of my investigations I have found it completely justified. The attractive features of the cosmic rays is their constancy.

They shower down on us throughout the whole 24 hours [We now known 64 million neutrinos go through a thumbnail every hour], *and if a plant is developed to use their power it will not require devices for storing energy as would be necessary with devices using wind, tide or sunlight. All of my investigations seem to point to the conclusion that they are small particles, each carrying so small a charge that we are justified in calling them neutrons. They move with great velocity, exceeding that of light.*

More than 25 years ago I began my efforts to harness the cosmic rays and I can now state that I have succeeded in operating a motive device by means of them. I will tell you in the most general way, the cosmic ray ionizes the air, setting free many charges ions and electrons. These charges are captured in a condenser which is made to discharge through the circuit of the motor. I have hopes of building my motor on a large scale, but circumstances have not been favorable to carrying out my plan.

Many indicate Nikola's conversion of a 1931 Pierce Arrow to run on "cosmic" rays was a hoax, simply because, one of the witnesses was Peter Savo, who said he was a "nephew" of Tesla, to reporter Derek Ahers in 1967, when his name didn't show up in an initial family tree attempt. Here is what Ahers said:

"Tesla took him to Buffalo, New York in 1931 and showed him a modified Pierce-Arrow car. Tesla, had removed the gasoline engine from the car and replaced it with a brushless AC electric motor that was powered by a "cosmic energy power receiver" consisting of a box measuring about 25 inches long by 10 inches wide by 6 inches high, containing 12 radio vacuum tubes and connected to a 6-foot-long antenna. Inside the box were some dozen vacuum tubes "70L7 type" and other electrical parts. Two wire leads ran from the box to a newly-installed 40 inch long, 30 inch diameter AC motor that replaced the gasoline engine. The

165

car was driven for about 50 miles at speeds of up to 90 mph during an eight-day period.

Tesla inserted the two metal rods and announced confidently, "We now have power" and then proceeded to drive the car for a week, "often at speeds of up to 90 mph." One account says the motor developed 1,800 rpm and got fairly hot when operating, requiring a cooling fan. The "converter" box is said to have generated enough electrical energy to also power the lights in a home. The car ended up on a farm, 20 miles outside of Buffalo, not far from Niagara Falls."

Tesla reportedly removed the box and returned to his New York City laboratory without revealing how he did it, and it would take until now for the feat to be reestablished by a man from Zimbabwe.

2015 Repeat

Sangulani Chikumbutso, dropped out of school had unseen visitors tell him what to do and he built a special receiving transformer that would repeat Tesla's experiment. His company, SAITH Technologies, developed and tested the Saith Fully Electric Vehicle (FEV) shown below. It has a top speed of 90 km/h and it gets its power from high frequency E-M frequencies just like Nikola did 80 years before. The E-M receiver can be removed and used to power a house just like Tesla stated. [See the last image- this man carries the R-F Receiver]

2016 Repeat

Toyohashi University of Technology and Taisei Corp built a similar E-M powered vehicle shown below left. Then they thought about using lower frequency E-M signals transmitted from the road and using coils in the tires for picking up the energy to continuously charge electric cars as shown middle with their factory shown right.

Piezo Vibration energy

Then we come to a somewhat removed technology that vibrated Piezo Crystals like the Pyramid electrical generator. The version on the left called "P-Eco smart car". Piezo-electric high frequency vibrating crystals in front and rear continuously charge the battery just like the second vehicle shown right. Certainly, the dozens of solar powered vehicle use a form of E-M waves to power vehicles as well.

167

Urine

I think we can trust that the "cosmic engine" probably was used to power a vehicle by Nikola just like he said, but let me diverge just a little. The urine car is substantially different and probably should not be in this list, but ever since the military urine batteries assured power in remote areas, conversion of urine to useable fuel has not only made an unusual vehicle where coffee drinkers can eliminate to activate, so to speak. One is shown next left, but I think they are going too far. In India, a soft drink made with bovine urine was recently under development and that country's leading Hindu cultural group hoped to market it as a "healthy" alternative to traditional soft drinks. It also is marketed as a detox drink.---Yuk!

 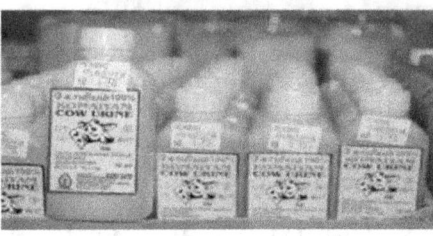 

1937 Project Rainbow

As all of Nikola's documentation was taken by the federal government within days of his death and some was held by the government as classified, we can believe any information about an error filled project called the "Rainbow Project" and its associated used of high frequency high voltage emissions to alter the environment would be of upmost interest to the details in this book. The focus of the experiment was on a small Destroyer Escort ship known as the USS Eldridge that modified time with some type of magnetic vortex.

The project to reduce the RADAR signature of ships was kept just as secret as the Manhattan Project that was trying to weaponize atomic fission. I know some of this will sound like a conspiracy, but just because something quacks like a duck, looks like a duck, and feels like a duck doesn't mean it's a duck. OK! It probably is a duck, but there are other possibilities so don't get caught up in those things.

The Eldridge-Massive Tesla coils and isolator coils were placed around the entire ship hull. Once in place, the high frequency, high voltage, magnetic field generators were excited and tested for a number of months in 1942. When the coils were all activated, the results appeared to be time travel and some associated spatial travel the government evidently became even more interested in Nikola's devices. A secondary project called the "Pegasus Project" would continue the time/spatial travel elements of this experiment later after Nikola passed away, but let's just concentrate on this experiment. No one really knows the full extent of what happened aboard the ship and many simply consider the whole Rainbow Project to be fantasy until the remarkable similarity of the work by researcher John Hutchison, using the same type equipment. The reason for the uncertainty is that people simply couldn't believe what happened in Philadelphia on board the USS Eldridge during a most unusual experiment COULD have happened. In subsequent experiments today, most notably by a Canadian named John Hutchison we find article that seem to disappear or be shifted to a different time and some items that change their crystalline structure or lose the effects of gravity. We will look at some of those experiments continuing Nikola's work, later, but right now, let's look at the pieced together details of the first large scale experiment into the magnetic effects of reshaping objects or concealing them.

While the Rainbow project was in full swing another important think-tank was doing the Manhattan Project. This group was trying to yank protons out of atomic clouds to make energy.

Like the Rainbow project, some don't believe the Manhattan Project ever happened either, but the cities of Nagasaki and Hiroshima both provide proof that yanking protons out of an unbelievably tiny atomic nucleus completely messes up the world near the proton dissection and the experiments did have the unbelievable results that were claimed.

Most of the reports of the Rainbow Experiment can't be 100% verified and initially they seem fanciful. The reports were of time and/or space dilation, and of the ship disappearing and reappearing at a different location, almost in the blink of an eye. Another sad result was the transfer of solid objects through solid objects. Unfortunately, the solid objects identified were human so there was a mess made during the experiment. With our new scientific knowledge, we might now begin to believe that there could be more to the story than we initially understood. Let's travel back to 1912 and see where all this came from.

Background-While there are many theories, this is one of the timelines that I like so I'm presenting it here. In 1912, a mathematician named David Hilbert developed several different methods and equations for something he called "Hilbert Space". Dr. Hilbert and another man, Dr. John von Neumann, a brilliant mathematician, got together in 1926 to expand the theories of Hilbert Space and time functions. The plot thickens as a Dr. Levinson came along and developed something he called the "Levinson Time Equations". All three of these guys began investigating the subject of invisibility in earnest in the early 1930's at the University of Chicago. Math was flowing all over the place and unbelievable answers were the beginnings of a new science of time dilation and magnetic field modification of molecular makeup. We can believe Nikola Tesla had read all the papers and had determined application using the ultra-high frequency and voltage, magnetic field generators he had used for many of his experiments.

Princeton University-By 1933, something called the Institute for Advanced Study was formed at Princeton University. The group doing this "Advanced Study" included Albert Einstein and good old Dr. John von Neumann. It is believed that the time dilation project was transferred to Princeton shortly thereafter and that the Rainbow Project was getting transformed from a Radar Cloak to something more amazing.

Of course, Einstein believed that there was a <u>single particle of vibrating nothingness</u> that made up everything, but, with combined theories of Tesla' vibrational matter concept and the others, one of the offshoots of the Relativity Theory was the generation of an incredibly intense electromagnetic fields around a huge object, like a ship, would cause the bending of light. Somehow this reaction would cause a time dilation

172

effect. Please remember, Tesla had believed that he had been transferred in time during one of his own experiments, so this was "old Hat" to him. It was explained as being like a massive spinning light refracting bubble. You may recognize the spinning light as the theory that Dr. Ron Mallett has concerning the design of a true time machine, but let me try to stay on subject. During these early days the new goal was to transfer ships to Japan without having to go across the Pacific Ocean. Once there, undetected, the War could be finished and the world would be saved.

Tesla-I'll tell you something right now. You could not do much electromagnetic craziness without getting Nikola Tesla involved to put things into perspective and let you know what the figures mean. He was the master of all vibrating electric fields. He understood, for instance, that if one could truly understand what magnetism really was he could control time itself. In 1936, the secret project was expanded to include Tesla as the director of the group for the most important 7 years of its mysterious life. Together, these smart guys studied the nature of relativity and invisibility. You can imagine what the experiments were like, but no record of the details has ever been published or even described outside the tiny group. Some indicate that with Tesla on board, partial invisibility was achieved before the end of his first year. Just as John Hutchison has done in his home in Canada.

Brooklyn Test-Certainly, details are sketchy about this whole super-secret group, but in 1940 a "full" test to make a ship disappear in our time was done in the Brooklyn Naval Yard. Another scientist named, Townsend Brown, evidently, became involved at this point. He had a background in gravity and magnetic mines so he fit right in with the others.

173

Test Number 2-By 1941, Tesla and his group were given a ship and coils were wrapped around the entire vessel to make it a huge magnetic coil. In some way, "Tesla coils" were also used as part of the experiment. Some say that this was not the actual USS Eldridge, but a different ship with the same designation so that there could be several ways to address deniability as no one really knew what was going to happen. I'm sure there were many miniature tests with vibrating invisible objects being transferred in some warehouse or basement, but the Navy needed to transfer ships so more testing would have to continue.

Oops!- Here goes the bad part. There must have been bad things happening in the smaller experiments. I suppose they tried transferring rats across a room and making them disappear and things like that, but there must have been issues with the crude results. The stories tell us that Tesla became very uncomfortable because he had a gut feel there would be problems with people being involved in the test. We are told that Tesla somehow knew that the mental state and bodies of the crew would be affected severely. He wanted more time to perfect the experiment but that Von Neumann character thought that there was a reasonable chance at success and the war needed Japan to be invaded. Tesla's requested more time and it is rumored [This is a rumor of a larger rumor if you are trying to keep up with the oddness of this story.] that Tesla went through the motions but secretly sabotaged an experiment in March of 1942. Because of this, he was either fired or quit the project.

Tesla Died-Within 10 months after leaving this special group, Tesla had announced to the world his massive discovery about the fabric of matter and the complete understanding of magnetics only to die before he could provide the information to anyone. Tesla died in 7 January 1943, a few months before the last and worst test that the "Rainbow Project" is best known.

Rainbow Project-As I mentioned above, at some point in time the studies became called "Project Rainbow". Allegedly, it was an experiment conducted upon a small destroyer escort ship, both in the Philadelphia Naval Yard and in a Norfolk Virginia Harbor location and the open sea between. Many indicate that the experiment was a complete success and an utter failure. The ship became invisible. According to some, the ship was seen one minute and it disappeared the next. In a fraction of a second the ship was seen in Norfolk, disappeared again and reappeared in Philadelphia. The bad thing is that the disappearing act was not the electromagnetic cloak or an easy to understand phenomenon. The ship, evidently had bent time or had bent space, or both, depending on how you wanted to view it.

Tesla's devices had turned the entire ship into a time machine. H.G. Wells would have been so proud!

Bad Experiment-If we piece the story together we get a reasonably clear picture of the strange experiment. On July 22nd, 1943, 7 months after Tesla had mysteriously died, the power to the Eldridge was turned on, and massive electromagnetic fields started to build up. Just think about huge tesla coils arcing and sparking and that might have been something like the effect, I suppose. A greenish fog slowly close over the ship and concealed it. When the fog dissipated, the *Eldridge* was gone. Everything seemed to have worked well above their expectations and about fifteen minutes later the greenish fog slowly reappeared, and the Eldridge was back. It had, evidently, made a travel through time. When the Eldridge was boarded, the crew topside was found to be disoriented and nauseous. It is quite likely they had no idea what had happened to them. According to some accounts, the crew was changed to a new crew and the equipment was modified to be less powerful. According to their cover story, the Navy wanted **radar** invisibility, and this complete disappearance was pretty scary. Certainly, the disappearing had already happened on a small scale in the lab, so they would have known about it, but secrecy was always a major objective during the war years.

The "Official" Navy Record-The Navy has admitted that the U.S.S. Eldridge took part in an "experiment". The experiment described is somewhat similar, but it does not admit the parts about the invisibility and things. The Naval report indicated that the test involved wrapping wire around the hull of the destroyer in an attempt to cancel out the magnetic fields of the metal on the ship.

Degaussing-This is known as degaussing and it is a common thing to do whenever electrons are bombarded on a surface as one would find in a CRT based Television set. Without the degaussing, there would be an electron buildup that would begin to distort the way electrons would hit the CRT so the images being made would look twisted. One could believe that by forcing the distortion of the electromagnetic fields associated with Radar, the radar returns would be distorted and make the return images seem to vanish in the background noise. This was a good cover story and even one of the ideas that probably had been focused on in the early experiments.

Elimination of Mine Detection-In the Rainbow Project details obtained since then, *the coils embedded in the ship would, supposedly, also render the ship "invisible" to underwater magnetic mines that rely on proximity sensors to trigger the detonation. These mine sensors operated by detecting magnetic fields around ships. Without the magnetic field, the ship would be able to pass through regions mined with these sensors, invisible to enemy mines, but not to radar or vision.* Let me tell you that the Navy does not always tell the whole story, but it is interesting that the admitted part included building a specially designed ship with a specially designed magnetic field generator to cause a level of invisibility. Let's see where the ship full of coils went next.

Worse Experiment- In October in 1943, a 2nd test on the Eldridge was performed. The electromagnetic field generators were turned on again, and the Eldridge became near-invisible. It was stated that only a faint outline of the hull remained visible in the water, but then the Eldridge vanished again. This time it was what they had really hoped for. Within a few seconds it reappeared in Norfolk, Virginia for a few minutes and disappeared only to reappear back at Philadelphia Naval Yard.

While the whole thing was pretty much Top Secret, it is said that the last experiment was accomplished while it was docked near the Merchant Marine ship S.S. Andrew Furuseth and one of its crewmen, Carlos Allende, wrote several strange letters to a Dr. Jessup, in the 1950's confirming the results. He was a tattletale for sure but there were a few other leaks to embellish what would become one of the most amazing tales ever told. The question still asked today is "How much of the story is factual and how much was embellished beyond the facts. We are just now just beginning to understand how probable the experimental results were and how the experiments, probably, showed us so much information about time travel.

Bad Stuff Again- After the last test most of the sailors were violently sick. Those were the lucky ones. Some of the crew was simply "missing". Some went crazy. A few of the crewmen had been partially teleported through a wall or deck. They became sort of "fused" with the ship itself. While this proved that people could indeed walk through walls, it was at a terrible price and absolutely no one knew exactly what had really happened. All the crewmen were either discharged with medical issues, had disappeared, or had died because of the experiment.

Much later, Einstein realized that each person has, essentially his own universe that he carries with him that is controlled in some way by his own consciousness. The effect of each person to the time warping would have been subject to that slight difference. Some may have returned a fraction of a second before the ship wall and "BAM" disaster. While they didn't understand the nuances of the time machine at that time, I'm sure they continued the studies.

It has been told that 2 guys, a Duncan Cameron Jr. and his brother, Edward, were in the control room to operate the equipment on the USS Eldridge. When everyone else had been killed or generally gone crazy, these two guys were shielded in the protected room. As they witnessed things falling apart, began shutting everything down, but it was too late. Some type of cover up ensued and the crazy sailors were placed in medical care as the United States went on to develop and deliver the Atomic Bombs and the outcome of this experiment almost disappeared from history.

What they had found is that the makeup of atoms and matter itself is very tenuous at best. It doesn't take much to disrupt and change them a little nudge with high frequency electromagnetic waves completely changed things. It apparently changed time itself for those in the vibrational field. Time was beginning to lose its secrets, but most people seemed to be unaware of the massive change that had taken place. As I mentioned before, one person, outside the government, that didn't ignore the implications was a man named John Hutchison. While his experiments certainly add credibility to Nikola's "rantings" let's first look at what Nikola had become. If you could see one of those science fiction characters with the huge head full of so much capability that we all looked like rats in comparison, which would be Tesla. A picture is shown next. He's the one on the top row.

The second row are artist depictions of alien pilots

I told you about the big head and I know it sounds like I'm rambling, but you just have to understand just how special this guy really was. Don't go thinking he was one of those aliens that were captured or seen leaving some spaceship. They supposedly have a big skull and small mouth similar to Tesla. I put a picture of the face of a reported "dead alien visitor" next to Tesla. I don't see the resemblance, but I must say he was mysterious. If those images aren't enough, the following group of deceased pilots of what are referred to as UFOs is shown next.

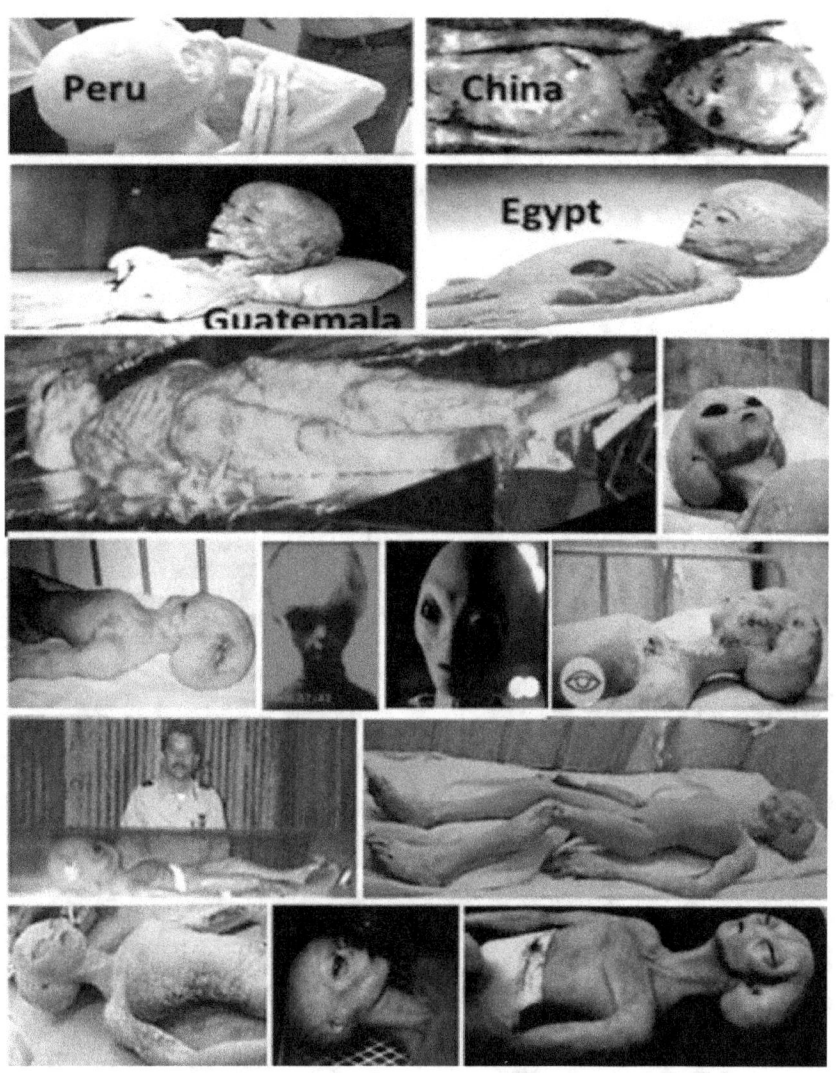

1943 Death and Legacy

Nikola Tesla died on January 7th, 1943 in the Hotel New Yorker, where he had lived for the last ten years of his life while he worked on the Rainbow Project. Room 3327 on the 33rd floor is the two-room suites he occupied. The remains of Nikola are tucked away in the little ball shown next and on display. The image to the right is Nikola not long before his death.

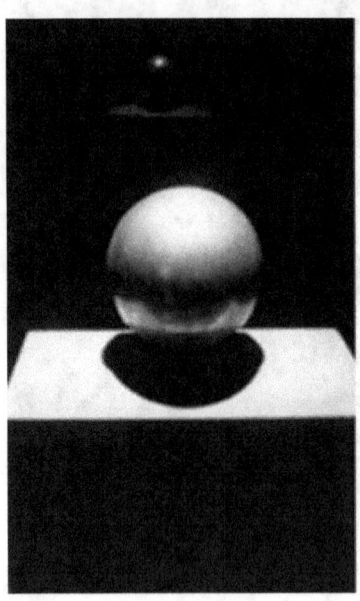 

During the night of January 7, 1943 Nikola Tesla died in his room at the Hotel New Yorker. While cause of death did not seem suspicious, timing is rather disturbing. The very next morning, after the inventor's death, his nephew, Sava Kosanovic, hurried to his uncle's room at the Hotel New Yorker. Let me tell you something about this guy as it might be important. He was an up-and-coming Yugoslav official with suspected <u>connections to the communist party</u> in his country. By the time he arrived, Tesla's body had already been removed. Kosanovic suspected that someone had already gone through his uncle's effects. He indicated technical papers were missing as well as a black notebook he knew Tesla kept. This was a notebook with several hundred pages, some of which were marked "Government." The following images are of his room a short time after his death and before the FBI came in and took everything.

A state funeral was held at St. John the Divine Cathedral in New York City. Telegrams of condolence were received from many notables, including the first lady Eleanor Roosevelt and Vice President Wallace. Over 2000 people attended, including several Nobel Laureates. He was cremated in Ardsley on the Hudson, New York.

In a speech presenting Tesla with the Edison medal, Vice President Behrend of the Institute of Electrical Engineers eloquently expressed the following: *"Were we to seize and*

183

eliminate from our industrial world the result of Mr. Tesla's work, the wheels of industry would cease to turn, our electric cars and trains would stop, our towns would be dark and our mills would be idle and dead. His name marks an epoch in the advance of electrical science." Mr. Behrend ended his speech with a paraphrase of Pope's lines on Newton: *"Nature and nature's laws lay hid by night. God said 'Let Tesla be' and all was light."*

Nikola Tesla's Awards and Recognition

In 1917, Tesla was awarded the Edison Medal, the most coveted electrical prize in the United States.

In 1960, Nikola Tesla's name was honored with an International Unit of Magnetic Flux Density called "Tesla."

In 1975, Tesla was inducted into the Inventor's Hall of Fame.

In 1976, the Nikola Tesla Award became one of the most distinguished honors presented by the Institute of Electrical Engineers. The award has been given annually since 1976.

In 1983, the United States Postal Service honored Tesla with a commemorative stamp.

In commemoration of his work in 1895, the Nikola Tesla Statue was made and is located on Goat Island to honor the man whose inventions were incorporated into the Niagara Falls Power Station. Tesla is known as the inventor of polyphase alternating current.

The Nikola Tesla Corner Sign, located at the intersection of 40th Street and 6th Avenue in Manhattan, is a constant reminder to all New Yorkers of the greatness of this genius.

Some of his books and articles are shown below.

 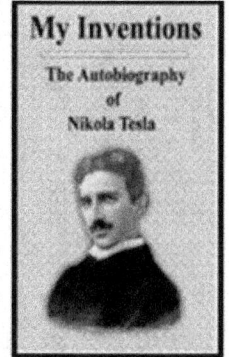 

Forgotten

Tesla's work almost fell into obscurity after his death, but since the 1990's, his reputation has experienced a comeback.

At his Death Site-We can believe they took all references to the Rainbow Project and anything associated with magnetic vortex for both levitation and time travel. According to museum officials at The Nikola Tesla museum in Belgrade, one of the things he left us were sketches of interplanetary ships. Before he died, Tesla, talked about a method of building a magnetic vortex to lift vehicles this way, *"No one knows what gravity is! Gravity must have spin. Diverging and converging rays cause standing waves in a unipolar generator."* Luckily one of his lab assistants, Otis Carr, continued in some of the work on levitation.

1955 Strange Case of Otis Carr

In 1955, **Otis T. Carr**, a lab assistant to <u>Nikola Tesla,</u> began a highly visible public effort to develop a prototype civilian spacecraft that could be mass produced in kits and sold to the public. The following images show a small model and a larger breadboard spacecraft.

The vehicle was to be powered by an electric generator drawing electrical energy from the environment, and would have produced an antigravity effect for propulsion. Carr claimed to have been taught all he knew about electromagnetic energy and antigravity principles by Nikola Tesla.

Carr's Description-Tesla's flying vehicle was to be powered by electrical energy drawn from the earth's atmosphere and stored in special coils. Frustrated by lack of industry support, Tesla revealed his radical ideas to the young <u>Carr</u> over a three-year period. Tesla taught Carr how electromagnetic energy could be freely harnessed from the abundant electrical energy in the atmosphere. Carr built a six-foot version to be tested for proving the feasibility of his ideas for a planned 45-foot prototype spacecraft. In November 1959 Carr successfully patented his design for a full scale civilian spacecraft he called <u>OTC-X1.</u> It had a circular design that made it look like a flying saucer as shown below. Please notice the strange pulsing magnetic capacitors [called Utrons] along the outer rim and a massive gyroscope with an odd core with perpendicular windings in its center.

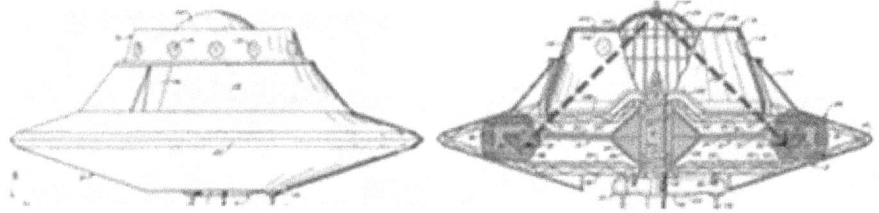

Carr described the Utrons as, "-*a storage cell for electrical energy. In operation it generates electricity at the same time it puts out electromotive force. This is the central power system for our space craft. The Utrons would supply a series of counter rotating magnets set in motion by the strange gyroscopic device with 4 windings at 90 degrees for one another. o move in one direction, less current would be*

supplied to one of the windings. The whole system was counter-rotating, the electro-magnets rotate in one direction and the accumulator/gyro rotated in another. The outer rim and capacitor plates were operated by a battery such that the windings rotated opposite to the gyroscope." The images below show close ups of the Utron devices.

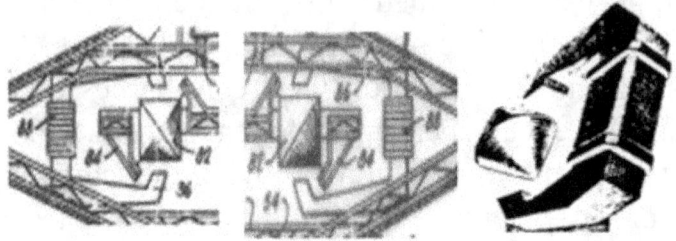

Below is a similar saucer shaped craft using the same principle as that described earlier. Please notice there were 12 Utron devices in the spinning base producing something called the Lorentz-O Vortex Tornado that actually pushes the vehicle into the air.

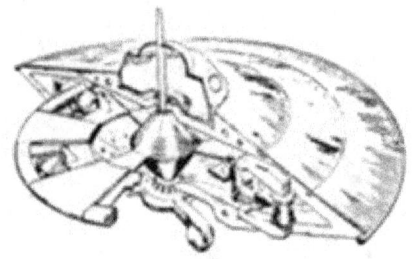
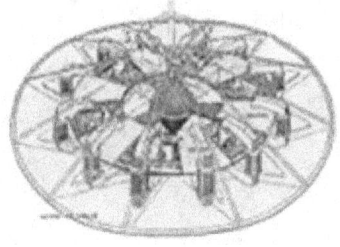

Carr indicated about his prototype *"We are able here, the first time to our knowledge, to use atmospheric electricity as a recharging system. This is done as a part of operational principal of the craft."* As far as witnessed testing, he demonstrated a small model in an interview as reported in *Fate Magazine*. He was able to finish his larger model and it was successfully piloted by Ralph Ring, but two weeks after the successful test of the OTC-X1. In 2006, Mr. Ring came forward and indicated Carr's operation was closed down by the FBI and other government agencies in a secret raid involving seven or eight truckloads of armed government personnel. FBI agents confiscated all the equipment including the OTC-XI prototype and then something strange happened soon afterwards. Ralph Ring went into hospital for a routine knee replacement operation. He accidentally received the wrong treatment, and nearly died three times. Now very frail and 71 years old, we may never know details of Tesla's spaceship legacy.

Later industry tests had mixed results. Below are the 1959 version attempted by NASA, left, and a more modern breadboard that has shown reasonable production of the Lorentz-O Vortex Tornado that could lift up the vehicle.

We will someday understand what Tesla had done to eliminate gravity for his flying machine. One man that has single handedly modified the minds of many with his widely viewed experiments to levitate heavy objects, make some disappear, and modify the molecular structure of others by using Nikola's devices. His name if Joh Hutchison.

1980 John Hutchison

Like the devices on the USS Eldridge, a rogue inventor has been using multiple high frequency radio transmitters and high voltage Tesla coils concentrated at a single location to produce "something" and when it is produced, things occur to objects in this "Field" just like they experienced in the earlier experiment. If you can transform a molecular structure, and tune it so that it can't see your body, you can simply walk right through the wall or walk through a block of time. In the case of the unfortunate sailors on board the Eldridge, they had no control over the modifications of the decks and walls they were near. John Hutchison has, almost accidentally, found the correct vibrations for particular elements of matter to make them appear invisible to one another as their "timeframes" did not stay in synchronization so that objects have passed through other objects without either being affected adversely. This same effect also has made objects appear to be gravitationally invisible to the earth to allow levitation. The vibrations also have mutated the materials to appear to be melted or jelly-like. John Hutchison is one of the miracle workers of this Age. He has discovered, almost by accident, what man has longed for over the centuries and Nikola Tesla had described 80 years before. While he, apparently, hasn't concentrated on time travel per se, his rediscoveries include levitation, transmutation, and invisibility. He was able to pass a piece of wood through a piece of metal. The reaction was halted and the wood remained in the metal. On another occasion, a knife became fused with a piece of aluminum when the electromagnetic field had been halted. Unlike other methods of forcing materials together the resulting combination of materials had no stress signs around the intruding material. Sometimes metals would appear to melt but surrounding wooden objects would not get hot as only the metal structure changed in the field. The pictures following show just a tiny amount of the oddness that Mr. Hutchison encountered. As snippets from a video, the first is a levitating bowling ball and pliers. These objects are invisible to the normal gravity.

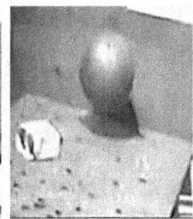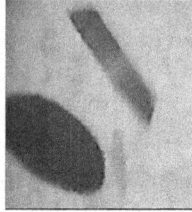

Then we see Aluminum changing into a jelly material.

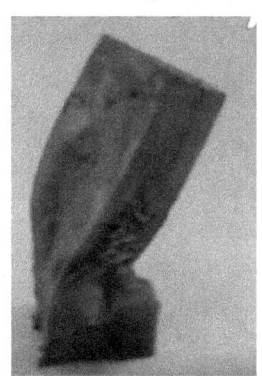

While sometimes the objects didn't levitate, they did change into different materials. Sometimes metals change

191

consistency and become jelly like before returning to metallic state in a stringy mess as shown next.

Objects Go Through Metal-This phenomenon starts really getting into the nuance of time travel. While it won't seem like it right now, I think later you will see how important this concept is to us traveling back in time. Sometimes, objects not only start floating, they go right through metal objects. If the process is halted before the material can go all the way through, the object traveling through the metal becomes part of the metal as shown next. The first picture is of a butter knife that began traveling through a metal plate. It ended up integrated with the metal. It should be noted that the crystalline structure of the metal was not affected by the intrusion. The second picture is of a piece of wood that was caught as it traveled through a plate of aluminum.

Invisibility-Now we are getting into something really important. Sometimes objects become begin to fade away and finally they become totally invisible. Below are a couple of images during the transition. Notice how you can see through the tubes.

Let me just put something out there right now. If matter becomes invisible and there is something we call conservation of matter, where does the matter go?

Keep this in mind as we try to get through time. It is important. That is all I'm saying right now. Please understand. If matter is changing, Time is changing as well. Both are critically linked in our current time-space continuum. I don't know how continuous this continuum thing is, but I like saying this phrase. What I mean is that matter doesn't really exist. It is simply the affect we witness as vibrating nothingness is linked by vibrational similarities to other vibrating nothingness, just like Nikola told us almost a century ago. Good old Einstein helped us understand how nothing was something and something was really nothing.

I'm certain that Mr. Hutchison is not the only one continuing the experiments. I'm certain governments have been investigating the time dilations experienced during the Rainbow Experiment for many years now and little time wrinkles have been made many, many times. We wouldn't notice them as each person sort of has his own universe and his own determination of time in that universe according to

Einstein. If you are going one speed and I'm going much, much slower, time for me is going much faster than it is for you and distance for me is extended substantially to you in your higher speed universe bubble.

Nikola Quotes

Nikola Tesla didn't just design things. On occasion, he would go out in public and say things. Here are a few that can help use.

Energy from Air--"*There is no more energy in matter than that received from the environment*"

Electricity is Everywhere- "*Electric power is everywhere present in unlimited quantities and can drive the world's machinery without the need for coal, oil, or gas.*"

Most Don't Thnk- "*I don't care that they stole my idea. I care that they don't have any of their own*"

Vibrational Matter- "*If you want to find the secrets of the universe, think in terms of energy, frequency and vibration.*"

On Intuition- "*The day science begins to study non-physical phenomena, it will make more progress in one decade than in all the previous centuries of its existence.*"

Future Women- "*But the female mind has demonstrated a capacity for all the mental acquirements and achievements of men, and as generations ensue that capacity will be expanded; the average woman will be as well educated as* *the average man, and then better educated, for the dormant faculties of her brain will be stimulated to an activity that will be all the more intense and powerful because of centuries of repose. Woman will ignore precedent and startle civilization with their progress."*

Tesla's Future- *"Let the future tell the truth, and evaluate each one according to his work and accomplishments. The present is theirs; the future, for which I have really worked, is mine"*

Discount Money- *"Money does not represent such a value as men have placed upon it. All my money has been invested into experiments with which I have made new discoveries enabling mankind to have a little easier life"*

Good and Bad of Man- *"Our virtues and our failings are inseparable, like force and matter. When they separate, man is no more"*

Research for Future Needs- *"The scientific man does not aim at an immediate result. He does not expect that his advanced ideas will be readily taken up. His work is like that of a planter -- for the future. His duty is to lay foundation of those who are to come and point the way."*

No Lover of Money- *"No desire for material advantages has animated me in all this work, though I hope, for the sake of the continuance of my labors, that these will soon follow, naturally, as a compensation for valuable services rendered to science and industry."*

Never Understand Life- *"Life is and will ever remain an equation incapable of solution, but it contains certain known factors."*

Service for mankind- *"The desire that guides me in all I do is the desire to harness the forces of nature to the service of mankind."*

God Made Man- *"Everyone should consider his body as a priceless gift from one whom he loves above all, a marvelous work of art, of indescribable beauty, and mystery beyond human conception, and so delicate that a word, a breath, a look, nay, a thought may injure it."*

Success-"*The practical, rational man, the observer, the man of business, he who reasons, <u>calculates, or determines in advance</u>, carefully applies his effort so that when coming into effect it will be in the direction of the movement, making it thus most efficient, and in this knowledge and ability lies the secret of his success.*

People Don't Take Care of Themselves-"*People could prolong their lives considerably if they would but make the effort. Human beings do so many things that pave the way to an early grave".*

Death Spawns Invention- *"If the genius of invention were to reveal to-morrow the secret of immortality, of eternal beauty and youth, for which all humanity is aching, the same inexorable agents which prevent a mass from changing suddenly its velocity would likewise resist the force of the new knowledge until time gradually modifies human thought."*

Become a Vegetarian-*On the general principles the raising of cattle as a means of providing food is objectionable, because, in the sense interpreted above, it must undoubtedly tend to the addition of mass of a "smaller velocity." It is certainly preferable to raise vegetables, and I think,* *therefore, that vegetarianism is a commendable departure from the established barbarous habit. That we can subsist on plant food and perform our work even to advantage is not a theory, but a well-demonstrated fact.*

Newness adds Sensation- *"We crave for new sensations but soon become indifferent to them. The wonders of yesterday are today common occurrences"*

Nikola also patented some of his ideas to try to keep others from stealing his work. While there were hundreds, let's just look at a few.

Major Nikola Patents

While Tesla had hundreds of patents all over the world and the United States. I have selected what I believe are the main ones to help those investigating Nikola. We are some are still classified so they are not listed.

U.S. Patent 0,334,823 - *Commutator for Dynamos* - 1886

U.S. Patent 0,335,786 - *Electric Arc lamp* - 1886

U.S. Patent 0,336,961 - *Regulator for dynamos* - 1886

U.S. Patent 0,359,748 - *Dynamo electric machine* - 1887

U.S. Patent 0,381,968 – *Electro-magnetic motor* - 1888

U.S. Patent 0,381,970 – Sys. of Electrical Distribution - 1888

U.S. Patent 0,382,279 - *Electro Magnetic Motor* - 1888

U.S. Patent 0,382,280 – *Elec. Transmission of Power* - 1888

U.S. Patent 0,396,121 – *Thermo-Magnetic Motor* - 1888

U.S. Patent 0,413,353 - *Obtaining DC from AC* - 1889

U.S. Patent 0,428,057 – *Pyro-Magneto Elec. Gen.* - 1890

U.S. Patent 0,433,702 – *Magnetic shield Primary* - 1890

U.S. Patent 0,447,920 - *Operating Arc-Lamps* - 1891

U.S. Patent 0,454,622 - patent on the Tesla coil - 1891

U.S. Patent 0,455,069 - *Electric Incandescent Lamp* - 1891

U.S. Patent 0,487,796 – *Elec. X-mission of Power* - 1892

U.S. Patent 0,511,559 – *Elec. X-mission of Power* - 1893

U.S. Patent 0,512,340 - *Electro-Magnets [bifilar]* - 1894

U.S. Patent 0,514,167 – *Coaxial cable* - 1894

U.S. Patent 0,514,972 - *Electric Railway System* - 1894

U.S. Patent 0,517,900 - *Steam Engine* - 1894

U.S. Patent 0,524,426 - *Electromagnetic Motor* - 1894

U.S. Patent 0,555,190 - *Alternating Motor* - 1896

U.S. Patent 0,567,818 – *Improved Capacitors* - 1896

U.S. Patent 0,568,177 - *Producing Ozone* - 1896

U.S. Patent 0,568,178 – *HF Wireless power* - 1896

U.S. Patent 0,568,179 - *Produce Currents of HF* - 1896

U.S. Patent 0,568,180 - *Electrical Currents of HF* - 1896

U.S. Patent 0,609,250 – *Elec. Igniter for Gas Engines* - 1898

U.S. Patent 0,613,809 - *Vehicles* by remote control- 1898

U.S. Patent 0,645,576 – *Power from balloon antenna* - 1900

U.S. Patent 0,649,621 - *Radio* - 1900

U.S. Patent 0,685,012 – *low temperature resonance* - 1900

U.S. Patent 0,685,953 – *Gnd conduction pwr & Data*- 1901

U.S. Patent 0,685,956 – *Resonance for gnd X-mission* - 1901

U.S. Patent 0,685,957 - *Ionization* & photoelectrics- 1901

U.S. Patent 0,725,605 – Wireless *Remote control* - 1903

U.S. Patent 0,787,412 – Resonant TX through Earth - 1905

U.S. Patent 1,061,142 - *Fluid Propulsion* - 1909

U.S. Patent 1,061,206 - Bladeless *Turbine* – 1909

CA135174 – *Tesla pump- Fluid Propulsion* - 1910

U.S. Patent 1,113,716 – large *Fountain little energy* - 1914

U.S. Patent 1,119,732 - *self-regen resonant Xformer*-1914

U.S. Patent 1,209,359 - *Speed-Indicator* - 1916

U.S. Patent 1,266,175 – *Improved Lightning rods* - 1918

U.S. Patent 1,274,816 - *Torque Speed Indicator* - 1918

U.S. Patent 1,314,718 - *Ship's Log* - 1919

U.S. Patent 1,329,559 – *Tesla Valve-Valvular Conduit* - 1920

U.S. Patent 1,365,547 - *Flow-Meter* - 1921

Canada B185446 - *Apparatus for Aerial Transport* - 1921

G.B. 186799 - *Balancing Rotating Machine Parts* - 1921

U.S. Patent 1,402,025 - *Frequency-Meter* – 1922

U.S. Patent 1,655,113 - *Aerial Transport VTOL* - 1928

U.S. Patent 1,655,114- *Dual counter rotating turbines to power the flying machine or spaceship-1928*

1932-*The Eternal Source of Energy of the Universe, Origin and Intensity of Cosmic Rays*, New York

1934-*Teleforce Particle beam weapon Art of Projecting* Non-dispersive energy through natural Medium gas focused, Directed Energy /Particle Beam

1937 Tesla stated *"It is not an experiment, I have built, demonstrated and used it. Only a little time will pass before I give it to the world."*

Conclusions

There is no question that some of the information in this book has been blown out of proportions and even established as fact without real proof, but one thing that can be said, Nikola was a great man. In addition to his AC system and coil, throughout his career, Tesla discovered, designed and developed ideas for a number of other important inventions—most of which were officially patented by other inventors—including dynamos (electrical generators similar to batteries) and the induction motor. He was also a pioneer in the discovery of radar technology, X-ray technology, remote control and the rotating magnetic field—the basis of most AC machinery.

With funding from a group of investors that included financial giant J. P. Morgan, in 1901 Tesla began work on the project in earnest, designing and building a lab with a power plant and a massive transmission tower on a site on Long Island, New York, that became known as Wardenclyffe. However, when doubts arose among his investors about the plausibility of Tesla's system and his rival, Guglielmo Marconi—with the financial support of Andrew Carnegie and Thomas Edison—continued to make great advances with his own radio technologies, Tesla had no choice but to abandon the project. The Wardenclyffe staff was laid off in 1906 and by 1915 the site had fallen into foreclosure. Two years later Tesla declared bankruptcy and the tower was blown up by the military, dismantled, and sold for scrap to help pay the debts he had accrued.

After suffering a nervous breakdown, Tesla eventually returned to work, primarily as a consultant. But as time went on, his ideas became progressively more outlandish and impractical. He also grew increasingly eccentric, devoting much of his time to the care of wild pigeons in New York City's parks. He even drew the attention of the FBI with his talk of building a powerful "death beam," which had received some interest from the Soviet Union during World World II.

Poor and reclusive, Nikola Tesla died on January 7, 1943, at the age of 86, in New York City, where he had lived for nearly 60 years. But the legacy of the work he left behind him lives on to this day.

Several books and films have highlighted Tesla's life and famous works, including *Nikola Tesla, The Genius Who Lit the World*, a documentary produced by the Tesla Memorial Society and the Nikola Tesla Museum in Belgrade, Serbia; and *The Secret of Nikola Tesla*, which stars Orson Welles as J. P. Morgan). And in the 2006 Christopher Nolan film *The Prestige*, Tesla was portrayed by rock star/actor David Bowie. In 1994, a street sign identifying "Nikola Tesla Corner" was installed near the site of his former New York City laboratory, at the intersection of 40th Street and 6th Avenue.

Before his death, Tesla had proclaimed to the world that he had finally uncovered all the secrets of magnetism. We would be interested because we will see that magnetism <u>may hold one of the keys to time travel so he may have found a SAFE way to exchange time reality.</u> Unfortunately, we may never know what he was talking about because he died a day or so after making the seemingly impossible claim. The reason that the claim seems impossible will be more apparent later, but right now let me say that magnetic vibration is now defined as <u>the essence of matter</u>. It can even be considered as matter itself, so Tesla sounded like he was bragging. Tesla usually didn't brag because he simply knew stuff, he said it was like information was pushed into his head. Anyway, when Tesla died, some person or persons came into his home and storage facilities and took all of his notes. I mean ALL his notes. Let me tell you how hard that probably was. This guy had created thousands of inventions, concepts, patents, and technical papers over decades. All of them were now gone.

As we start to look for insight with respect to the greatest inventor of our time, I think we first need to read a quote by Nikola himself. He stated *"It was like I could see the Past, Present, and Future at the same time."* Many of his discoveries he gave credit to images he received in a state of semi-consciousness or even by images of people who helped guide and instruct him. With the aid of these seemingly crazy-man helpers, he solved problems of motor commutation, AC electricity, participatory anthropic description, vertical flight, plasma ribbons for lighting, particle acceleration, manufacture of lightning and through-air electrical distribution, magnetic time distortion, Earth resonance for transfer of electricity, cosmic energy harnessing, understanding of the Aether and vibrational matter, free-energy from differential thermal environments, Astro-physics and talking to the heavens, energy from the ionosphere and magnetosphere, the secret association of Magnetism and gravity, and many more truths we are only now finding out were not crazy ideas.

Who were the his Invisible Friends?

I want to talk about these unseen "helpers" just a minute and see where they came from or at least present one possibility. For this I need to go to the book of Genesis that talks about the *"Giants of Old, before Adam"*. Certainly, scientists have found hundreds of shoeprints and modern footprints integrated with dinosaur prints along ancient beaches, skeletal remains of giant people that are either petrified or crumble when disturbed; ancient physical evidence of electricity, nuclear power, growing stones, high levels of manufacturing, flying machines, genetic science; and hundreds of ancient texts describing a more ancient race of people who were cursed by God such that when they died, they had a secondary "life" as something called demons. A tiny sampling of ancient, giant feet and shoe prints in stone are shown below integrated with giant dinosaur prints. The guy that wore the shoe on the far right, second row, stomped on two trilobites and killed them before they became extinct.

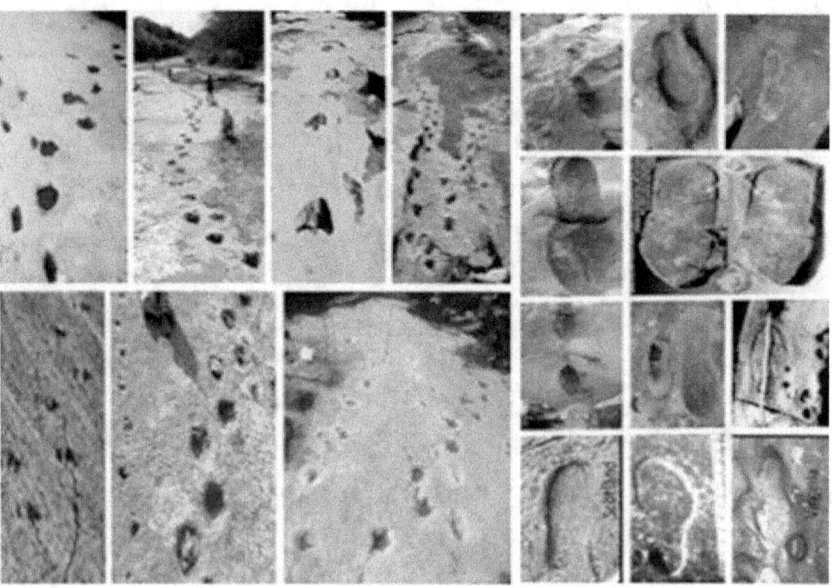

While most demons would possess humans to regain a glimpse of reality, we can believe, so would wish to do whatever they could to regain favor with the creator. These would become teachers to special people who had an empathic nature. Nikola Tesla had the characteristics that seemed reasonable. Trauma had pushed him to the limit of sanity, his unusual photographic memory allowed much of his research to be conducted completely in his head, and he was obsessed with gaining understanding to the point of pulling him away from the world, sexual ambiance, power and money hunger, the idea that only the "now" is real, and those things which could cloud one's mind for going beyond reality. I know this sounds creepy and strange, but what if he and some of the others who saw future events and or described future concepts; Nostradamus, Mother Shipton, Dr. Cayce, and others, were actually given insights by these demonic souls desperate to make amends for past evils. By the way; the Greeks did not have the complete horror of demons and thought of many as teachers. In religion of Ekankar, people are aided by unseen "teachers". Maybe Tesla's insightful helpers were real. Let me reinitiate fear. Tesla's "demons" were not normal. Don't mess with demons. Most are only out for destruction of mankind and they will do anything to accomplish that desire.

How Crazy Was Nikola?

I'm not saying Nikola was not crazy. One thing that seems to be pretty evident, whenever one is possessed, it messes with the mind. Nikola aligned his beans before eating, couldn't look at pearls, couldn't touch hair; he could only eat items in quantities of 3, and he talked to pigeons. Even with all of that, he was so focused on his work, nothing kept him from learning, testing, and understanding the secrets of our universe. OK! When he almost lost his eye, he did slow down on X-Ray investigations but that was minor compared to his push for understanding and helping humanity. Some of his insights are just now being understood. Let's review just a few in case I did not describe their significance adequately.

- **Relativity**-When he said, Einstein's Theory of relativity was flawed because many things go faster than the speed of light, people scoffed until they found that even electrons spin around a nucleus faster than the speed of light and are still in our "reality".
- **Particle Beams**-When he indicated he could make a particle Beam that could kill miles away, people scoffed, and now we have them.
- **Drones**-When he said Robotics boats and planes would fight our next wars and even showed us how it could be done, people scoffed and now we do it.
- **Vibrational Matter**- When he said "matter" didn't exist except as a vibrating Aether, people scoffed, and now all physics is passed on this important concept that is the basis for all modern physics.
- **Resonance**-When he said we could use the earth to transfer electrical energy by finding it resonance, people scoffed and many still hold onto outdated "laws of earth resistance, but some are now finding this can be done.
- **Aliens**-When he said we could hear messages from space and life on planets. They scoffed some more, but now major industries are set up to test evaluate, and understand other worlds. Nostradamus, the Egyptian King Thoth both described a major encounter between us and those not

living on Earth in the future and now people think they may have known something.

- **Levitation**-When he said, he could control gravity by controlling magnetic fields they scoffed and now many are building levitating crafts with new knowledge.
- **Solar Power**- When he stated the sun held all the power we would need for everything, people scoffed, but now Solar energy
- **Consciousness Control**- When he stated human thought could control the Earth, stars, and the universe; people scoffed, not remembering the words of God incarnate who told his followers, "--with "faith" one can move entire mountains by simply telling them to move. Today the theories of Anthropics and Quantum Mechanics both rely on cognizant viewers as "reality controllers".
- **Energy from the Air**- When he indicated he could collect energy form the air for distribution; people scoffed. Today, a man from Zimbabwe has made a multi-resonant transformer and receiver that can pick up all types of electro-magnetic frequencies in the air. With this energy we can now power a car or even part of a home. The first two are RF powered cars by Saith Technologies. The third image shows the RF energy receiver/charging system. Just as an aside, Sangulani, founder of Saith Technologies, indicated he was given insight into the design by an unseen teacher.

Next group are form Toyohashi University with Toyota. The first two are vehicle models and the last image shows the RF energy collection system used to power the motor.

- **Vertcle Take-Off and Land Vehicle**-When he indicated one could take off vertically for air transportation. He was thought of as a silly man, but now even VTOL Jets are being used, Helcopters are used around the world, and now helicopter style drones are everywhere. All from his insight.

- **Flourescence**- Somehow Nikola knew the high-frequency, high-voltage Electricity could initiate a glowing plasma just like what happened when the Earth and Venus neutralized their voltage differences. Now we use plasmas for all types of things including fourescent lights, welding, cutting, etc.

- **Magnetic-Gravity**-When he dared to indicated Gravity was only a symptom of mass spin rather than a feature of the mass itself, they scoffed. Now we are finding that changing the spin changes the gravity of any mass.

On and on we can go, Nikola, seems to already know about the things that would make our life more livable. We can believe all of these things were commonplace, thousands of years ago, and we are realizing these characteristics of our environment much quicker this time thanks to Nikola Tesla.

Thanks for reading and stay safe.

About the Author

Steve Preston is a long lime author of scientific, esoteric facts. His books focus on the painful truths rather than whitewashed details that make us comfortable. If you are interested in the truth instead of comfort, please review other works by Mr. Preston as shown below. The images are some from Egypt taking the older version of a taxi. To the right the writer is shown in the Jewish Negev desert of Israel where the Dead Sea Scrolls were found.

To the left below are a couple of pictures as we searched the New Zealand caves searching for ancient Maori artifacts and the last image is of the author investigating the Acropolis concerning ancient Athens Greece.

Development of Mankind

The First Creation of Man-book 1 History of mankind
The Second Creation of Man-book 2 History of mankind
The Creation of Adam and Eve-book 3 History of mankind
The Antediluvian War Years-book 4 History of mankind
Man After The Flood-book 5 History of mankind
Close Look at Ancient History-book 6 History of mankind
A New View of Modern History-book 7 History of mankind
The Twentieth Century and Beyond- Book 8 History of Mankind

Through the Bible Series

Abraham to Moses-Second part of the Bible
Adam to Abraham- First Part of the Bible
Moses to Jesus- Third part of the Bible Series
Understanding the New Testament-4th part of the Bible Series

Bible History, Correction, and Analysis

Adam's First Wife-Story of Lilith
Closer Look At Genesis- 200 ancient text confirm Genesis
Exploring Exodus- Reviewing the Details of "Exodus"
Errors in Understanding- Interpretations of the Bible
Expanded Genesis- Apocrypha and other Jewish texts
Exploring Genesis- Reviewing the details of "Genesis'
Incarnations of God- How often did God become Incarnated?
History Confirmed By The Bible- Science confirms the Bible
New look at the Bible- Questions in Interpretation
Old Testament Used By Jesus- Ancient Jewish texts
Why the King James Bible Failed- Issues with KJB
Differences in the King James Bible-Added Apocrypha

Anomalies

Religious Anomalies-Explaining unexplained scripture
US History Errors- Explaining disregarded events and causes
Space Anomalies- Explaining anomalies of our near planets
DNA Anomalies- Explaining hidden truth tested by DNA
Modern Misconceptions-Examining Anomalies in Science
Anomalies in Flight- Strange Flying things

Ancient Technology

Titan Gods- History of the Ancient Giant/gods
Kingdoms Before the Flood- Pleistocene humans

Current Events and Fears

Strange, Powerful, Dangerous Women- Women of History
Allah' God of the Moon- Terror of Muslims
American School Disaster- fear in our country
Make Your Own Global Warming
Monsters are Alive- Post Pleistocene Monsters
Terror of Global Warming- Fake issue uncovered
The Antichrist- Many demonic possessed rulers
The Devil- Of Demons and their master
Vampires among Us- How Demons and Vampires are similar
Humans on Display- Slavery and Human Zoos

New Look at Physics

Anthropic Reality- We control our Reality
Consensus Science- Fake Science
Complex Earth- Truth behind Earth's development
Is Time Travel Possible? Science of Time Travel
Retiming the Earth- Eliminate of Nuclear Decay Errors
Releasing Your Consciousness- Beyond our SELF
Slip Through a Wall- How to walk through solids
Our 12-Dimensional Universe- New science of our Universe
Mystery of Photons and Light- Science of Photons
Of Heaven and Hell- scientific descriptions
Meaning of Life and Light- Detains of New Science
Vibrational Matter- New Science of Quantum Fluctuations
Does Science Confirm the Bible? - Application of Physics
When Earth Exploded- Near Collision with Mars

New Look at Biology and Self

Understand Your Heart- New Discoveries of the Heart-brain
DNA of Our Ancestors- Tracing DNA of ancient man
God Didn't Make The Ape- New science on ape Evolution
Lizard People- Mutated People of the Bharata War
Creation and Death of Dinosaurs- Why Dinosaurs died
Races of Men- Tracing DNA of Humans
Tracing Cro-Magnon to Jesus- Follow new findings
Self, Soul, Spirit- Three components of Life
Self-Virtualization- New science of reality
True Happiness- Self Actualism and Beyond
Life Resonance- Unusual capabilities of men

Awaken the Departed- We can talk to the Dead
Biophotonics and Healing- How Photonics used in medicine
Homo-Erectus as a Man- Characteristics of Homo-Erectus types